Essex Place-Names

First published 2007

by Historical Publications Ltd
32 Ellington Street, London N7 8PL
(Tel: 020 7607 1628)

ISBN 978-1-905286-21-8
British Library Cataloguing-in-Publication Data
A catalogue record for this book is available from the British Library

Typeset in Book Antiqua by Historical Publications Ltd
Reproduction by Tintern Graphics
Printed by Edelvives, Zaragoza, Spain

The Illustrations

The following illustrations were reproduced by kind permission:
Chelmsford Museum: *28*
Colchester Museums: *74, 85, 97*
A. Corder-Birch: *84*
Essex Journal: *66, 85, 86, 87, 102, 105, 106*
Essex Record Office*: 3, 9, 10, 32, 53, 55, 58, 59, 61, 63, 64, 72, 83, 89, 91, 100, 104*
Essex Society for Archaeology & History: 9, 20, 22, 33, 45, 54, 60, 69, 79, 80, 82, 101
A. W. Fox: *86, 87*
Maldon Town Hall: *102*
Valence House Museum: *112*
P. Waterman: *42*
Nos. *23, 25* and *52* are by courtesy of the owners of the properties
Nos. *90, 98, 105* and *108* were supplied by the publisher.
All other pictures were supplied by the author

The front cover illustration is of Ingatestone High Street, from the north in the 19th century; on the back cover is one of the many attractive village signs in the county. Both were supplied by the Author

Essex Place-Names

Places, Streets and People

James Kemble

HISTORICAL PUBLICATIONS

Contents

List of Illustrations

Preface

"We know what we belong to, where we come from
and where we are going, not in our brains but with our roots".
Noel Coward, 1899-1973.

Few people are wholly disinterested in the locality where they live and something of what went on there before the present. One of the key ways of finding out how our surroundings have developed over the centuries is from learning how and why names have become attached to towns, villages, streets, roads, buildings, fields and woods. Yet the discovery of the original meaning of our place-names is often fraught with difficulty and it is a brave man (or a foolish one) who will insist on certainty. As quipped Oscar Wilde "Don't say you agree with me. When people agree with me I always feel I must be wrong."

This publication has been written in response to many people's increasing interest in their local surroundings and its history. The Volunteer Recorders, who are the backbone of the Essex Place-names Project, have been researching and recording place- and field-names as well as personal names from old maps and documents, and investigating their derivations and origins. The local Recorders are especially well placed to be able to relate place-names to the topography and to investigate on-the-ground the significance of obscure names.

The founding of the English Place-name Society in 1923 brought together several academic institutions for the study of place-names and has done much to foster a systematic collection of these names county by county. Dr. Percy Reaney wrote the Society's Essex Place-names volume collection in 1935 and it remains an invaluable resource; it is however in need of some revision and updating as further documentary sources have come to light, been studied and interpretations revised. On a national as opposed to county scale, Professor Eilert Ekwall's *Oxford Dictionary of English Place-names*, 1960, has been followed by Victor Watts' *Cambridge Dictionary of English Place-names*, 2004.

The active involvement of the Essex County Archivist Ken Hall and Heritage Conservation Manager David Buckley was crucial to the initial Essex Place-names Project set-up, and those at the Heritage Conservation Section, Owen Bedwin, Paul Gilman, Alison Bennet, Sally Gale, Maria Medlycott and at the Essex Record Office, Richard Harris, Stuart Warburton, Jill Baker and the searchroom staff have willingly continued their essential support, as has the Essex Society for Archaeology and History.

I am grateful to Dr. C.R. Hart for access to some of his unpublished research material and for permission to use it freely, and to Drs. Margaret Gelling, Richard Coates and David Parsons of the English Place-names Society for their support and interest. It has been a pleasant cooperation with Mr. John Richardson of Historical Publications Ltd., who has professionally piloted the production of the book.

PREFACE

I hope that this book may act as an introduction to a fascinating aspect of our history and, more particularly, as a spur to further enquiry. While I acknowledge with gratitude the advice and assistance I have received from those whose whole-time business is Etymology and Place-name Study, any errors are mine, but, as again quipped Oscar Wilde, "don't shoot the pianist, he's doing his best". The searchable database is on the internet at www.essex.ac.uk/history/esah/essexplacenames and in the individual parish booklets published by and obtainable from the Essex Record Office. Further information is available from the Essex Place-names Project Coordinator, Essex Record Office, Wharf Road, Chelmsford CM2 6YT.

James Kemble

The Development of Place-Names

"Bright with names that men remember, loud with names that men forget".
Charles Swinburne, 1837-1909.

While most of us take for granted the names of places we live in, travel through or visit, a second thought may cause us to ask how those places originally got their names. Because names change over centuries, to try to understand their meaning it is necessary to go back to the earliest recorded spelling we can find in old documents. These early spellings may reveal much more than the origin of the name, often a record of the landscape at that time, the use to which our predecessors were putting the land; they reflect invasions and settlement by outsiders. Probably over 90% of the town and village names we now have originated before 1500. Even the new post-war conurbations (such as Basildon) take their origin from the small settlement which was named before the 11th century.

It is possible that just a scattering of still-existent names was first used by those who inhabited the county in the Stone and Bronze Ages 3000 or more years ago. They are likely to denote landscape features which the settlers would need to refer to, like rivers and hills. Colne is a possible example of such a name which was passed to the Iron Age people who settled from around the 8th century BC.

Celtic was the unwritten language of the tribes which inhabited Britain from about 700 BC; its derivatives are spoken now in Wales, Scotland, Ireland, and, until the 18th century, in Cornwall. A few names still in use probably originated with the Iron Age Britons who spoke Celtic before the Romans came in 43AD. Places such as Leyton include the name of the river Lea which is probably a Celtic-British name. River names such as Pant, Thames and Stour are probably British or Old European names, meaning something like 'running water', 'dark waters' and 'strong', recorded first using the Roman alphabet. 'Dovercourt' contains the Celtic-British element dubr- meaning 'water' or 'stream' plus Latin cohors, an enclosed yard, or possibly 'land cut-off'. Some pre-Roman British river names such as Hyle and Beadewan are recognisable only by associated places which grew up beside them, Ilford and Baddow, the river name having been replaced by Roding and Chelmer *(Fig. 1)*.

Although Latin was the language of the military and administration from the 1st century AD, Latin names surviving in Essex are few, suggesting they were not persistently used or that the Romans were content to use the existing British names. Such 'Roman' place-names are found in copies of documents such as the Antonine Itinerary (AI), a list of stopping-places for travellers throughout the Roman Empire, the *Notitia Dignitatum*, a record of the forts along the British and continental coasts, and of authors such as Julius Caesar and Tacitus.

The failure of Roman names (Caesaromagus (Chelmsford), Canonio (Kelvedon), Durolito (?Chigwell)) to survive may be testimony to the perception by the native Briton, albeit under a Roman administration, that attaching names to their hamlets and farms served little useful purpose. A few were adopted into Old

1. Little Baddow mill house, built c.1900, on the River Chelmer.

English, such as those containing 'chester' and 'wic' derived from Latin castra and vicus meaning 'camp' and 'minor administrative centre' (Chesterford, Wickham Bishops). One instance of a Saxon personal name affixed to a Latin loan-word, fontana, survives in Tolleshunt (documented in AD 1000 as Tollesfuntan, Toll's spring or well). Strata, paved, was loaned becoming 'stræt' and 'street'. It is likely that Latin further infiltrated into England from the church clerics who continued to use Latin (and adopted Greek words) for their writings from the 5th century. Place-names with a Latin element such as Tilbury-juxta-Clare, Beaumont-cum-Moze, Wendens Ambo are survivals of medieval Latin.

The name of the county Essex is first documented in the 8th century by Bede, the priest historian working at Jarrow in Tyne and Wear. Writing in Latin he calls it regno Orientalium Saxonum, 'to the region of the East Saxons'. It included London. The Saxons had migrated from the Low Countries and northern Germany in the 4th, 5th and 6th centuries, speaking a Germanic language, now called Old English.

Old English names form the great majority of the place-names still in use today. There persist in Essex a few place-names probably from the early Saxon immigration and colonisation phases (5-6th centuries AD) which suggest that the native British retained their own identities in some settlements. Thus Walton may be the tun, settlement, of the wealas, serfs/Britons, and Cumberton in Little Chishall may be the denu, valley, of the Cumbra, the Britons. Walden, first recorded in Domesday Book (DB), is 'valley of the Britons', Saffron being added by the 16th century from the crop being grown here. Walter Hall in Boreham, Walhfare in 1062, 'passage of the Britons', suggests a British enclave. Also from this

2. All Saints Norman church, Vange.

3. A railway cut through Bartlow Romano-British mounds, 1864.

early phase are those names containing -gau, district, such as is contained in Vange, fen district, *(Fig. 2)* and Ginges (Ingatestone, Mountnessing, Ingrave etc), 'people of the district'.

Names with -ham, settlement (e.g. Burnham, settlement by a stream) probably are earlier than -tun names (Dunton, hill farm). Lack of clarity in the spellings of the earliest documentation we have between -ham and -hamm, enclosure often by water (as in East and West Ham), can be a cause of confusion. The very common -ing suffix is the result of it being added to another word to form a place-name. Thus we have Clavering, the place where clover grows. But -ing can also conceal an earlier spelling -ingas, so Nazeing, territory of the people on the spur, or Roding, territory of the followers of Hrotha. There is evidence that people from Essex settled in other parts of the country. In Hampshire is Exton (in 940, East Seaxnatune, farm of the East Saxons) and in Cambridgeshire Saxton, settlement of the Saxons.

Place-names descriptive of the lie of the land, the topography, are numerous. The Anglo-Saxons had a rich vocabulary for features such as hills, valleys and countryside. Thus 'beorg', later borough, berg or berry, is round-topped hill (thus Bergholt), 'dun', later down, is wide flat-topped upland (thus Downham and Dunton), 'hlaw', later low, is tumulus, often man-made (thus Harlow, Boblow and Bartlow *(Fig. 3)*, and 'hoh', later hoe, is hill-spur or promontory (thus Holland and Langenhoe). Old English 'denu', later -don or -den, (not to be confused with 'denn', woodland pasture), is wide sloping valley (as Debden and Theydon), 'halh', later hale or hall, is a nook or separated land (so Strethall, Chrishall, Coggeshall, Chignall and Willingale), 'slæd', later slade, is wet flat valley (thus Fulfen Slade south of Saffron Walden).

The Viking invasions of Britain from Norway and Denmark in the 9th and 10th centuries had an influence on Essex place-names mainly in the north-east Tendring peninsula. -by names (with exceptions derived from Old English -bury, fort, and byge, bend), such as Kirby are almost certainly Scandinavian. But while only a tiny scattering of such names was recognised (for example Thorrington and Frowik) until only a few years ago, increasingly more, such as Moze, Chich (St. Osyth) and Tendring itself are being suggested as of Scandinavian origin as scholarship progresses. In a few instances a Scandinavian place-name has superseded an Old English one, such as Whitby formerly Streoneshalh, 'Hviti's farm' once 'Streon's corner of land', an example of landholder displacement? In some instances a Scandinavian personal name has been used as a place-name, thus Oskethey in Halstead, Asketill's enclosure, and Skighaugh in Great Oakley, Skegge's enclosure.

That there are many similarities between Old English and Scandinavian makes it difficult to be sure from which a name derives. Thus either Old English fugol or Scandinavian fugl (bird) could have formed Fulenesse, Foulness, though the paucity of Scandinavian names in the south of the county makes the former more likely. But such paucity could be held to be due to a failure to recognise Scandinavian names in the south rather than Old English names, and thus to put the argument in danger of becoming circular. There are words 'adopted' from Scandinavian into English such as bigging, an outhouse, Great Biggins (Finchingfield), and holmr, island in marsh, Thornholm (East Ham). There are also names which originate from a family whose origins may have been Scandinavian but who had settled in England for many years; Archesdana, Arkesden, Arcel's valley, may be such an example.

Of similar interest to and perhaps more accessible for research by the enthusiastic layman not versed in Old English or Celtic are the place-names from the 16th century up till the present. These are found in estate maps, Rentals, Court and Assize Rolls, Surveys etc. which are, on the whole, not too difficult to read; many have been published in transcripts. These contain evidence of social history, agricultural and industrial usage, such as Ozier, Saffron, Tenter and Kiln field, for willow and saffron-growing, cloth-, brick- and pottery-working. 'Camp' field-names (such as Campers Mead in Woodham Ferrers) may refer to where a type of football was played, though earlier the term seems to have referred to uncultivated or enclosed land.

To the archaeologist, place- and field-names hold significant clues to sites of past activity when now little may be visible above ground. Thus 'barrow' and 'low' names may be former sites of tumuli, 'borough' and 'bury' names to fortified sites (and hence, later, to manor-houses), 'street' to a paved road, possibly of Roman origin. Sites of pagan worship or believed to be the work of the devil or dwellings of spirits may have acquired names such as Grimsditch (in Mountnessing) and Hobgoblers (Woodham Ferrers). Boundaries, often delineated by banks and/or ditches, may be represented by dyke names, such as Mar Dyke, boundary ditch, the boundary between the Hundreds of Chafford and Barstable. Other linear banks may also have acquired 'dyke' names; the system of banks west and south of Colchester which began in the Iron Age include Gryme's Dyke, Sheepen Dyke, Lexden Dyke *(Fig. 4)*, etc.

Sites of buried settlement may be present under fields such as 'Blacklands' (Cressing, Woodham Ferrers) and 'Blackleys' (Great Leighs), the black soil representing settlement decay. 'Wickham' names such as Wickham Ley in Little Waltham may be sites of Roman settlement, often close to a Roman road.

4. Lexden Dyke, Colchester.

'Tunstall', 'Dunstall', 'Tunstead' and 'Dunstead' names (for example in Mundon, Downham and West Hanningfield), originating from Old English tun-stall and tun-stede, farmstead, may be places of now lost farm settlements.

The study is also of interest to family historians since many personal names, owners, tenants, parties to agreements etc. are incorporated into place-names. In the 11th century many names were affixed 'de', of, such-and-such a place, thus William de Warenne, from Varenne, Seine Inferièure. Later English place-names such as Dagworth in Suffolk give origin to the name of the family of Richard de Daggewurth. The dropping of 'de' continued into the 15th century. Hugh Naylyngherst no doubt came from Naylinghurst by Braintree, and William Hekford of Great Bentley from Heckford in Stanway.

Contrarywise, some place-names take origin from personal names, though this category seems on the whole to be more ancient. The meanings of many Saxon personal names are unknown, though a few seem to have been nicknames (such as Hwita, white, and Crispa, curly). Thus Whitchurch could be 'church that was coloured white, i.e. lime-washed', but Whitburn (County Durham) 'Hwita's tumulus'. In Essex Witham, the form unchanged since Domesday, may be from Old English wiht, curve, but one early form gives Hwitham suggesting the possibility of 'white' or 'Hwita'. When a Saxon personal name has not been found documented independently, the origin of a place-name from a person must be an inference, and pre-Conquest personal names which have a proven connection to a place-name are rather few, mainly because of our limited documentary sources. (Wimbish is a possible instance, from Lustwine, also known as Wine, who left the manor to Ely monastery). From the 12th century onwards we are on surer ground with place-names such as North Weald Bassett, Aythorpe Roding and Sible Hedingham.

'Collective' names also give origin to place-names. Bishops, earls and knights, who more likely owned rather than lived on these estates, give names such as Wickham Bishops (of St. Paul's, London), Earls Colne (earls of Oxford) and Knighton Wood in Woodford and Fryerning (knights and friars of St. John of Jerusalem). Abbots, monks and hermits are remembered in Abbots in Lawford (the property of St. John's Abbey Colchester), Monks Wood in Little Hallingbury (of Bermondsey Priory), and Bedeman's Berg in Writtle.

The Saxon Influence

Evidence from documentary sources and archaeology is of an increasing Anglo-Saxon presence in Britain by migration from the Low Countries and northern Germany from the 4th century onwards. With the exception of the civil and military administration who spoke Latin, most of the native population were speaking the British (Celtic) tongue, but increasingly this was replaced by Old English.

The early Saxon settlers concentrated along the eastern seaboard. Evidence of their presence has been found at Mucking, Shoebury, Tilbury and Maldon Heybridge, though some penetrated inland as is shown by a Saxon cemetery at Great Chesterford and houses at Rivenhall. Place-name evidence suggests that huge areas of Essex became settled by Saxon communities, some taking their name from that of the tribe or tribal leader. Dengie, 60000 acres, the central area around Ingatestone, Ingrave and Mountnessing named Ginge, 18000 acres, and The Rodings, 12000 acres constitute huge tracts of land apparently under a unified influence. It has been suggested that such regions may have been vacated by the native British in the 5th and 6th centuries through plague and migration, leaving them open to the new settlers, though much settlement is likely to have been by integration and assimulation. Almost certainly there will have been a period when both British and Old English were spoken, by some bilingually.

Incoming Saxons heard existing British names and adapted them (without necessarily understanding their meaning) into their own tongue. This led to some tautologies; thus Penhill in Devon combines British for 'hill' pen plus Old English hyll. So far there have not been Essex examples securely identified, suggesting the transition to English was relatively quick. The Saxons modified some of the Celtic place-names by hybridising the British name with an English affix. Thus 'Colchester' added ceaster (fort) to the British river-name Colne. Chatham Green near Great Waltham may be British cet, wood, plus Old English ham(m), settlement or corner.

Around AD 590 comes evidence for the emergence of kingship in Essex, almost certainly arising out of a system of petty chieftains or filling a power vacuum. The recent discovery of a kingly or princely burial at Prittlewell, Southend dated to around 620-640 is evidence of such 'royal' chieftainship. Another fine princely burial was discovered at Boreham near Chelmsford. It is possible that at least some of these tracts represent land-grants to petty chiefs and their followers of already roughly-defined territories which had become depopulated, and were therefore "spare and available" without the need to displace significant British populations.

The Saxon Charters, mostly records of land holdings and transfers (which for Essex start in the 7th century), Writs (letters of administrative business) and Wills constitute the bulk of the early sources of place-names which have come down to the present day. These are sometimes the originals but more often later copies. Since there was much incentive to lay claim to land, which are authentic and which forgeries has taxed the judgement of Saxon historians. Those Charters referring to the monastery at Westminster are especially suspect but may contain some elements of truth.

Saxon personal names originally had meanings, thus Godgifu meant 'good gift' and Mæssa 'glorious victory'. These have become place-names, Good Easter and Mashbury, to which the original meaning no longer necessarily applied. While there is pre-Norman Conquest documentary evidence for some of these personal names, for many there is no such sure evidence, and their once-existence is surmised. Names that are found in our main source, Domesday Book, compiled in 1086, of course contain Old English names that existed before then. There are post-Conquest records which strongly suggest that such Saxon personal names did exist. Thus in 1166 a woman named Pacchild lived in Essex, and this name probably explains the Domesday manor Patching Hall (in Broomfield) and Paglesham.

The late Saxon kings such as Alfred and Æthelstan appointed ealdormen to adminster their lands, in Alfred's reign over a shire but by the 10th century over larger swathes; one ealdorman controlled East Anglia, and possibly one covering Essex, Kent, Sussex and Surrey. These men were extremely powerful and exercised authority on the king's behalf (and their own), witnessing charters. They gave council to the king, formulated laws, in the king's absence acted on his behalf and were responsible for raising the king's army. Men like Brihtnoth led the Saxon army against the Danes at Maldon and was killed there in 991. Thegns, a word originally meaning 'one who serves', attended the king or their ealdorman lord. Though often not specifically called so, many of the "free men" of Domeday Book holding land before 1066 were probably thegns. The Danish kings, starting with Cnut in 1016, introduced the class of jarl or eorl, whose role and function were virtually indistinguishable from the ealdorman. Earl and bishop sat jointly as presidents of the shire court. In 1017 the Danish eorl Thorkell the Tall controlled Essex and East Anglia. Earldoms of naturally hereditary and powerful families, such as the Godwines of Wessex of whom King Harold was one, emerged. Earldoms could however be stripped by the king and not until the 12th century did earls have jurisdiction over a single shire, Geoffrey de Mandeville, son of the Domesday Geoffrey, being the first earl of Essex.

The names of ealdormen, earls, thegns and their families became attached to the places they controlled. Thus a woman called Ælfgyth , who may or may not be the same as the wife of Atsere Swearte who gave Wennington to Westminster Abbey before 1042, may have given her name to Aveley. Likewise Godgifu, wife of an ealdorman, gave her name (after the Conquest) to Good Easter, that part of Easter which was in her possession. It is possible that the name of Lustwine, grandson of ealdorman Brihtnoth, is contained in the first element of Wimbish, an estate of which he willed to Ely monastery *c.*1036.

The Imposition of Norman and Anglo-Norman names

The Norman landlords who accompanied William the Conqueror in 1066 and their successors have imprinted significant legacies not just of church architecture and furniture *(Fig. 5)* but of vocabulary and names, sometimes as an affix to the Saxon place-name, thus Tolleshunt D'Arcy from the landlord Robert Darcy. Theydon Garnon's landlord was the family of Ralph Gernun (Old French grenon 'moustached'), descendant of the Domesday holder Robert. Poor Park in Wethersfield takes its name from William Lepuiher ('a Picard'), and Mark Hall in Latton from Adelulf de Merc (from Marck, Pas-de-Calais). Some are nick-names like Tolleshunt Malger (Major), Old French for 'council-prober', or Viel Engayne (Colne Engayne), Viel the Tricky. The Norman custom of adding a post-fix to their name (e.g. Ralph de Limésy) permits us to pinpoint their origin in France, and some of these who gained Essex estates are shown in *Fig. 6*. The custom continued in England but by about 1200 many of the aristocratic families considered themselves assimulated Englois. Though William's conquest is often referred to as 'Norman', it will be appreciated from how wide an area his followers came, from Flemings in the east to Bretons in the west, with Normandy centered on the archdiocese of Rouen (Ouen).

The great Survey carried out at King William's command in 1086 entailed the visit by his commissioners to the shires to record what every landholder owned in land and livestock. For the eastern counties their base was probably Bury St. Edmunds. They recorded the names of manors and Hundreds (collections of manors). For Essex it is fortunate that we possess an early record, known as Little Domesday Book, before it was abridged and condensed as happened with most of the remainder of the country. Little

5. The Norman font at Fyfield.

Domesday records, in Latin, not only the new Norman landholder and tenant but also often the Saxon holder or tenant and the size of his holding before 1066. These Saxon names sometimes provide a link with the Saxon Charters, Writs and Wills. The place-names are of course those in use in the Saxon period, occasionally adapted a little by the Norman clerk who may not have fully understood Old English. They contain Old English elements, such as dun, hill and denu, valley, stede, place and treow, tree. It is in Little Domesday Book that the great majority of the earliest documentation of the names of manors, later to become villages, survives.

No doubt it became necessary to distinguish one Saxon manor from another after the Conquest, so Woodham Mortimer became distinguished from Woodham Ferrers and from Woodham Walter, all, before 1066, just Wuduham. Not all such affixes are directly from France; in Domesday Great Easton was Eistanes, but the manorial hall became Eystan Blamuster, a contraction of Blau minster from the family who came from Whitchurch in Shropshire.

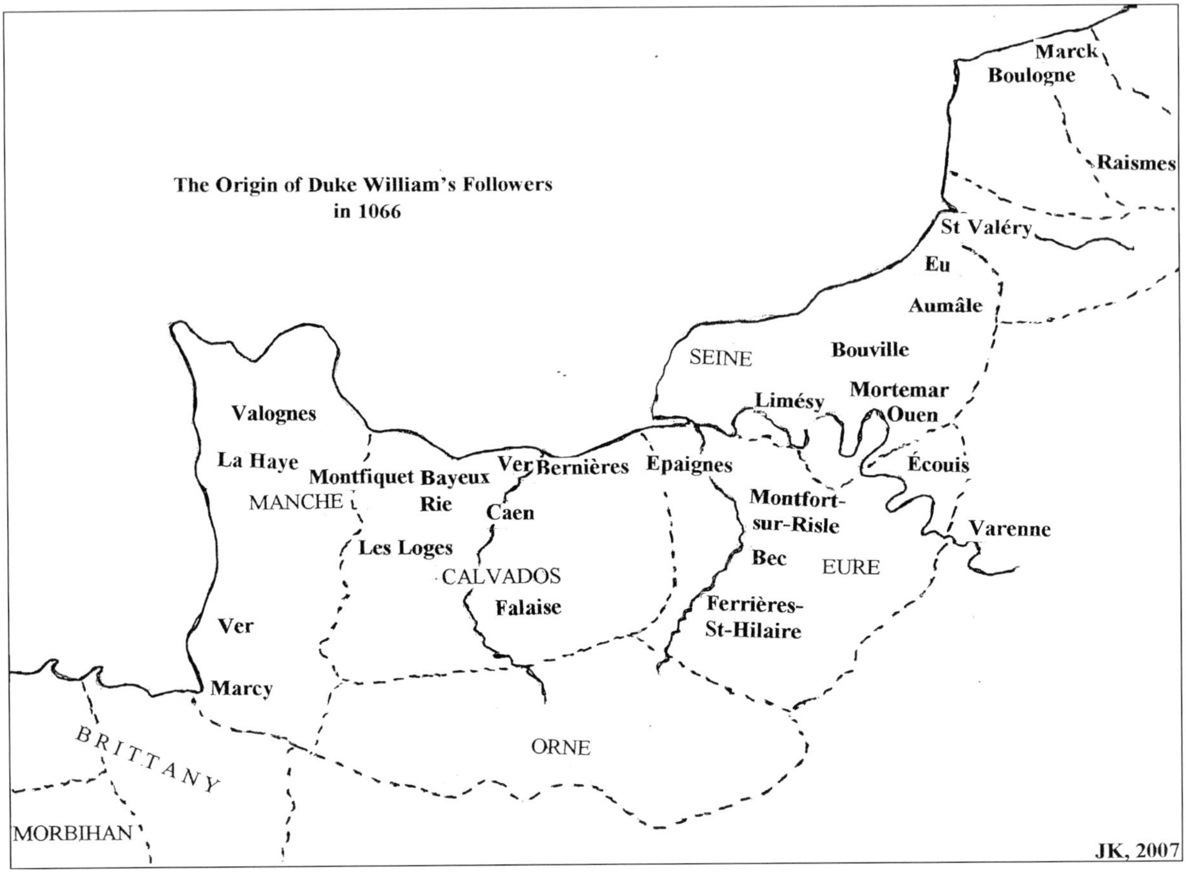

6. The origins of Duke William's followers in 1066.

Purely French names are rare compared with the Saxon; Pleshey is a direct adoption from Norman-French plaisseis, a fence of living hedge, Bealmont replaced Fulanpettae for Beaumont-cum-Moze. It is clear that very few of the Saxon landholders in possession in 1066 were still so at the time of Domesday 20 years later; almost all had had their land confiscated for the new Norman barons. Of three monasteries existing in 1066, Waltham Holy Cross, Barking and West Mersey priory, a cell of Rouen (Ouen) Abbey, all continued to hold lands in 1086. French monasteries and nunneries such as St. Valéry and Caen benefited from grants of Essex manors.

In Essex, later medieval landholders' Anglo-Norman personal names have sometimes replaced earlier Saxon habitation names. Thus Bigods (in Great Dunmow) from the family of Bartholomew le Bigod in the 13th century was, in Domesday Book Alferestuna, Aelfhere's farm, and Domesday Wigghepet, Wicga's hollow, became Rokell's Farm (in Elmdon) from Humphrey from La Rochelle.

The predominance of French-speaking Normans in position of influence in the 11th, 12th and 13th centuries significantly influenced the Old English spoken by the native Saxon majority. Many documents were written in Latin, the language of the church from which words have probably been loaned directly ('chester' from castra, 'port' from portus), but the intrusion of French vocabulary into Old English was significant. But since the Normans had difficulty pronouncing some sounds, substitutions have occured, thus Old English Burgefen ('fortification in marshland') became Bulphan. The English initial 'th' sound resulted in Old English 'thornen dun' becoming Torinduna, 'hill with thorn-trees', then losing the initial T to Horndon when the vill was divided into East and West.

Most of the administration was carried out in Norman French, soon Anglo-Norman. After the loss of Normandy to the English crown in 1204, English began to reassert itself as a language of the influential. In 1362 Parliament decreed that lawsuits could be conducted in English but not until 1399 did Henry IV become the first ruler of England after The Conquest whose mother tongue was English. Henry V (1413-1422), sensing the anti-French feeling in England weary of the Hundred Years' War, decreed that letters from the Palace of Westminster should no longer use French but English. It was the Masters of Chancery, the office responsible for the records required for administering the kingdom, who recognised the need to standardise spelling so that documents could be understood everywhere in the land. It was these Masters and a few Senior Clerks in the 15th century who made decisions about which spelling should be adopted as 'standard'. The spellings which were adopted were mostly drawn from London and the Midlands, apparently not following consistent 'rules' consequently leaving us with 'mouse' and 'mice' but 'house' and 'houses', 'he' and 'his' but 'she' and 'hers'.

By the start of the 15th century English was the tongue in which teaching in schools was conducted. In particular, inflectional endings of words were lost, leaving but a few such as the 's of the possessive (father's), 'ox, ox-en', or vowel change, 'foot/feet' in the plural. From Scandinavia were introduced 'kirk' from kirkja and 'earl' from jarl, the Saxon 'ealdorman' now obsolete. From an amalgamation of Old English with Anglo-Norman French influenced by the Scandinavian elements slowly emerged Middle English, the transition virtually complete by the time Geoffrey Chaucer (1340-1400) was writing *Canterbury Tales*. By then "English" had acquired parallel words from different sources, thus from Old English sheep, deer and help, from Old French mutton, venison and assist. It may here be noted that the modern practice of trying to suggest archaism by signs such as 'Ye Olde Shoppe' disguises that the 'Y' is merely a substitute for the now-disused Old English letter thorn, written þ, which represents the sound 'th', not 'y'.

Influence of Paganism and the Church

There persist in place-names some remnants of the early pagan influences on the land to which Roman writers occasionally refer. Pliny writing in the mid-first century AD observed the Druids' custom of holding oak groves sacred and juice of mistletoe being drunk to enhance the fertility of bull sacrifice. Water, springs and wells were places of ritual, providing a communication with the underworld. Bogs, ponds and marshes were sites of ceremonies, and cult activities around pools have been excavated at Witham and Harlow.

Local beliefs in dwarfs, goblins, demons and sprites, kongon, puca, and scucca are remembered in Quindal Hole (Stebbing), Hobgoblins Hole (Great Baddow), Puck Lane (Waltham Holy Cross), and Shackmore Heath (Lawford). Nickerlands in Stanford Rivers and Nikermadwe in Greenstead-by-Colchester are place-name references to water-sprites, and fairies maybe inhabited Fairy Field in Braintree. Precisely when these beliefs originated is difficult to say, though some may be of prehistoric, or at least of later pagan, origin. Gransmore Green recalls the Old English poem (which took its form *c*.8th century) *Beowulf*, the heroic warrior who slew the monster Grendel.

Though Christianity had been brought, probably by the Roman military and traders, to Britain by the 3rd century, the Anglo-Saxon immigrants of the 5th and 6th centuries were pagan. They brought with them their gods such as Woden (called Grim), Thor and Thunor. These have left their names in Wodnesfeld in Widdington, Grimsey Grove in Fingringhoe, Grymes Dyke in Greenstead (though our earliest documentation is later than the Saxon period) and Thundersley. Pagan sites may be alluded to by names such as Swanstead field in Stisted (swine's head), and perhaps ruins of pre-Roman or Romano-British sanctuaries are remembered in Harrowdown in Birdbrook (Old English hearg, heathen temple), Willeys field in South Weald (weoh, idol). The Old English word hlaw usually evolving into 'low' often signifies a mound or hill, sometimes of burial. Thus Mutlow next to Audley End railway station was a tumulus from which Saxon shield bosses and a spear were excavated, almost certainly from a Saxon grave.

7. The 7th-century chapel at Bradwell Othona.

By the 7th century, Christianity was becoming more widely accepted, at least by the elite. About 653, Cedd, a priest from Northumbria came to Essex at the invitation of the East Saxon king Sigeberht to convert his people. He set up churches at Tilbury and Bradwell, the Saxon chapel which still survives *(Fig. 7)*. Some time after this a monastery church, the Southminster, was established south of Cedd's minster. On high ground above the Ingrebourne river, Upminster was built, perhaps marked now by the church which dates from the 12th century dedicated to the early 7th century bishop of Canterbury St. Lawrence *(Fig. 66)*. Tradition holds that Saints Cedd and Chad preached at Chafford (Chad's ford) in the 7th century. Priests who lived at these minsters taught in hamlets

8. Thoby Priory Tudor House and 14th-century church in 1818.

and open fields where they erected preaching crosses. These places are remembered in names such as Maldon, 'hill with a cross'. Holy places where people assembled were designated by -stow, thus Walthamstow, 'the holy place of welcome'.

Gifts of land to religious institutions such as monasteries, abbeys and priories were seen as indulgencies to buy prayers for the departed's soul. Gifts to the monks can sometimes be differentiated from those to the bishop or abbot. As early as the beginning of the 8th century land in Dengie was given to the bishop of London at St. Paul's, but separation of monks' and the bishop's land seems to have taken place by a century later. By the time of Domesday their respective holdings are identified separately. So, *c.*998, we find Hinawicun ('the community's' Wickham) i.e. Wickham St. Paul belonged to the canons, but Wickham Bishops was held by the bishop. Stapleford Abbots belonged to Bury St. Edmunds at Domesday but the earliest documentation of the affix 'Abbots' is found in the 13th century, and Domesday is careful to differentiate the abbot's holdings from St. Edmund's community.

St. Paul's, as the seat of the Bishop of London which formerly lay within the land of the East Saxons and where there has been a church at least since the first bishop Mellitus in the early 7th century, has been a prominent institutional Essex landholder. Its estates can be traced in some detail from the 10th century and earlier. One of the earliest, Tillingham, was an endowment to St. Paul's *c.*610. It owned property in regio Deningei *c.*709 which was probably serviced by the monks at Southminster while those at the church at Bradwell Othona served Tillingham.

Subsidiary churches probably were built first by lords of the manor in the 10th, 11th and 12th centuries. For their own convenience rather than that of their tenants, they often sited them near the manor house. Hornchurch 'the church with horns' is documented in the 13th century as is Berechurch, 'the church constructed of boards'. Several place-names derive from the dedication of the church, so St. Lawrence in the north

Dengie peninsula, and Magdalen Laver, combining St. Mary Magdalen to the earlier Lagefare.

The two religious Crusader Knight organisations, the Knights Templar and the Knights Hospitaller of St. John, had significant land-holdings in Essex. These started as military-aid organisations for pilgrims to the Holy Land and both became wealthy and politically powerful, to such an extent that in 1312 the former was banned and some of its possessions were awarded to the latter. Temple North Field in Cressing and Temple Wood in West Hanningfield were possessions of the former, while the Hospitallers held Fryerns in Basildon and Fryers in Little Sampford. St. Bartholomew's Hospital in London, part of the priory founded in 1123 whose church and cloisters can still be seen, owned Hospital Shaw in Aveley and Friers in Hatfield Broad Oak, while St. Giles' Hospital, Maldon, held Spital Mead.

Some priories' (and even some priors') names still persist in present place-names, despite the monasteries' Dissolution by Henry VIII in the first part of the 16th century. Thus Latton Priory persists in Priory Farm, Colne Priory owned Monks field in Aythorpe Roding and Thoby Lane perpetuates Tobias, the first abbot of Thoby Priory founded in the 12th century in Mountnessing *(Fig. 8)*. Other now-defunct monasteries, priories and nunneries, are recorded in place-names such as Monksfield (in Toppesfield), Friars field (Black Notley) and Mince Croft (North Weald Bassett), myncen = nun.

The Evolution of Place-names

*"The very names of things beloved are dear,
And sounds will gather beauty from their sense".*
Robert Bridges,1844-1930.

The earliest documentations of place-names which have survived to our time are often linked to the personal name of an owner, bequestor, grantor or recipient of an estate. Thus we read that Leofstan granted land in Southchurch to the monks at Christ Church Canterbury (in 823) and that King Edgar sold Navestock to St. Paul's (in 867). The dating of such charters is the result of careful comparison of clerical styles, forms of wording, of lists of witnesses to the charter and their known years of influence, the order in which their names appear in seniority, etc. Often only an earliest and latest possible date can be ascribed.

It may be possible to follow the ownership of estates, particularly when we have a series of Wills and Charters. Thus at Barking, the earliest document copied from one which dates *c*.666 refers to Berecingas et Beddanham. Bede, writing *c*.735, refers to monasterio Bericinensi. The Will of ealdorman Ælfgar which has been dated *c*.950 left land at Baythorn to his daughters whom he names, their children and, if none, to Sce Marie Stowe at Berkynge, i.e. Barking monastery (CS 1012). His elder daughter (who married King Edmund) left in her Will dated between 962 and 991 an estate at Woodham to Sca Marian cyrcan (church) at Byorcingan to her sister Ælfflæd who in turn willed it to St. Mary's, Beorcingan. In Domesday Book, Sce Marie de Berchingis held 30 hides of Berchingas from which Jocelyn Lorimer had taken 24 acres after 1066. Thus over 400 years we trace the name essentially unchanged except for the early loss of Beddanham. The 'modern' spelling Barking does not appear until the late 13th century, but clearly is the same name. Neither the Baythorn nor the Woodham estate were with Barking monastery in 1086; all three Woodhams and Baythorn were held by Normans.

Ownership by a monastery was often long-lasting over many centuries. In 867, St. Paul's monastery purchased 15 houses at Naesingstoc (Navestock) from King Ethelred at the request of the Bishop of London Deorwulf for 60 gold mancuses. After *c*.998 and before 1066, St. Paul's lost this holding and William I in a charter with that date restored it to the monastery, referring to an unjust dispossession. The canons held Nasestoca and Nessetocha in 1086. Navestock is documented in the 14th century, and at least part of the parish was St. Paul's at the Dissolution in 1544, a documented ownership of nearly 700 years.

The phonology of English is not the purpose of this book, but the change of sounds of vowels and consonants over time has been studied in some depth (Jackson, 1971). Sound changes in languages of Indo-European origin (from which Celtic, Old English, Norman French and Old Scandinavian all descend) have been shown to follow similar patterns. Spelling was phonetic and no doubt local accents, not to mention dialects, would have perplexed some of the Domesday scribes as they took down evidence from the local Essex witnesses. Changes resulting from the Norman Conquest and the difficulty of pronouncing some Old English sounds has already been alluded to, but variant spellings may give clues to origins which would be absent if only one spelling existed.

Prepositions and adjectives such as 'at', 'by', 'under', etc. may become attached to a feature to form a place-name. Thus, Noak Hill was atten ac, at the oak-tree, with the 'n' becoming attached to the 'ac'. While the Old English for 'island' is eg, Ray Island, West Mersea was 'æt thære eg' ('at the low-lying land'). (It will also be appreciated that 'Mersey Island' is a tautology, 'Island-by-the-Pool Island'). Similarly, the Mynster became the Upmynstre, on higher ground than the one on lower ground at Barking.

Old English was an inflexional language, that is it consisted of verbs and nouns to which endings were added to indicate person, tense, gender and case. Middle English showed a progressive loss of these endings. Many place-names were formed from the dative case, 'to'. Sometimes as we have seen the preposition was written, but increasingly this was omitted leaving the dative noun to stand for the place. Thus, in the 8th century, Bede writes for Barking 'in Bercingum', dative plural, by the 10th century just 'Byorcyngan'. Ashen preserves the dative 'æt thæm æscum', at the ash-trees, the 'n' substituted for 'm'. In the same way, the genitive case 'of' also was used to form place-names. Thus 'bulene mera' became Bulmer, pool of the bullocks, and 'brocheseued' became Broxted, head of the badger.

Words often evolve by shortening, and place-names follow this trend. This may be a loss of letters or a whole syllable (thus Mitteslea became Mistley, Lalleforde became Lawford). A vowel may shorten. Thus a long 'a' is preserved in Stane Street, but shortened in Stanway, Stanwægun. Such vowel shortening when it occurs in a terminal vowel has resulted in its disappearance altogether: Deppedana from 'deopan dene' became Debden, (at the) deep valley.

Examples of consonant substitution such as 'n' to 'm' have already been mentioned (Sanforda, sandy ford, to Sampford, and Fanbruge, fenn-bridge, to Fambridge), but other examples include 'd' and 'th' to 't' (Hadfelda, Hethfeld, heathy open country, to Hatfield), 'c' to 'g' (Paclesham, settlement of Pæccel, to Paglesham), 'sc' to 'sh' (Scopingeland, land with sheds, to Shopland), 'f' to 'v' (Fengge, marsh-district, to Vange), and 'v' to 'f' (Vithleris to Fithler's Farm, Writtle).

A Classification of Place-names

Place-names may be allotted into a number of categories and such classification assists study of their definition and origins:

1. Habitation names.
Old English ham and tun imply settlement, enclosure, then developing into a group of dwellings. Our present word 'hamlet' has a similar root. Names with wicham such as Wickham St. Paul very frequently lie near to Roman roads and wic here suggests an administrative settlement deriving from Latin vicus. Later wic names apply to village, port, saltworks, street or farm (often dairy-farm as in Chiswick Hall, Chrishall, 'cheese-farm'). Also implying habitation are worth, stede and thorp, as in Bobbingworth, Stisted and Easthorpe.

2. Topographical or geographical names.
The Anglo-Saxons, much more aware of their natural surroundings than the present generation of mainly town and city-dwellers are, had a rich vocabulary for natural features. While we might use 'hill' to denote any elevation, they used beorg, dun, ofer and scylf to denote different shapes of elevation so that a traveller could identify a hill from its configuration. Streams (Old English burna or broc, burn or brook), fords and bridges give their names to places, as Stambourne, Stanford and Stambridge. Trees figure often, as Elmstead, Ashdon and Boxted. How land was used is denoted by feld, leah, wudu, open country, wooded clearing thence pasture (-ley), and woodland, thus Shenfield 'fine open country', Bromley 'broom wood/clearing', Woodham 'settlement in the wood'.

9. Gosfield Hall, built c. 1550 by Sir John Wentworth, c. 1820.

3. Personal and Folk-names. Settlers from the continent brought their tribal names with them, sometimes their leader's name. These names are qualified as place-names by the addition of -ing or -ingas, thus Bocking and Roding. A few place-names in the north-east of the county are probably Scandinavian. Tendring may originate from a Danish place Tønder in Jutland, implying 'people from the place where tinder could be gathered'.

4. Animal and bird names. Domesticated as well as wild animals and birds figure in place-names. Thus (High and Good) Easter 'sheep-fold', Foxearth, Pewit Island (in Great Oakley), Faulkbourne (falcon?) and Gosfield (goose) *(Fig. 9)*, but not Finchingfield which was in Domesday Book Fincingefelda, the open country of Finc's people, underlining the importance of early spellings.

5. Tree and vegetation names. Names such as Oakley, Notley (nut), Saffron Walden, Ramsey (wild-garlic?) and Cressing (water-cress) were probably named from the predominant trees in the vicinity and the crop.

6. Back-formations. Some more recent place-names denote places which were once differently named and take their newer name from one nearby. This applies, for instance, to the rivers Brain, Roach and Ching, formerly Aqua de Wyham, Walfestreme and Boorne, but now back-formations from Braintree, Rochford and Chingford.

7. Genuinely new names. Finally a few place-names are apparently new creations, though even these tend to incorporate older elements. Thus Castle Point merges Hadleigh *Castle* with Canvey *Point* to form the name of the new administrative unit of 1974.

Field and Minor Place-names

"What's in a name? that which we call a rose
by any other name would smell as sweet".
William Shakespeare, 1564-1616.

Field-names which include minor place-names are of particular interest in that they provide a way of reconstructing local past landscapes and industries. The Tithe Awards of *c.*1840 and Estate maps are a rich source of these names which can sometimes be traced back into earlier documents to 13th and 14th centuries or even earlier *(Fig. 10)*. For many parishes in Essex the abandonment of the former Open Field System in which large open fields, often called North Field, East Field etc, were subdivided into smaller fields, probably occurred by the 12th and 13th centuries when new names for these smaller units were required. These names are preserved in Feets of Fine (property exchange records), Manor Court Rolls, Cartularies of religious institutions, etc. From the 16th and 17th centuries, landlords had their estates mapped and often the surveyor included the names of the fields and the acreages on the map, even sometimes recording the type of crop.

It has become clear that many of these field-names are of great antiquity, little change being needed in an essentially conservative occupation of agriculture which, until the agricultural revolution in farming practices in the late 19th and 20th centuries, was altered little since the Middle Ages. The smaller fields with their hedgelines which were so common until the Second World War and were held to typify 'English countryside' were remarkably persistent until the 20th century but, particularly during the middle of that century reversion to huge prairie-like fields (with loss of field-names) which better suited heavy mechanisation became again the norm especially in the eastern part of the county. It is therefore the Tithe Awards and Maps dating from the middle of the 19th century usually with a host of field-names that provide the main county-wide corpus of 'recent' names. Occasionally fields with names such as West Field, North Field persist into the 19th century and when these are not in parishes where Enclosure occured late, they are remnant evidence of the former Open Field System.

These Open Fields were divided into Furlongs or Shots by pegs to mark each villager's alloted part of the field for a year. These Furlongs were named and those names have often been carried into the later Enclosed Field-name. The 3 or 4 large open fields were 'rotated' between crops, one being left fallow on which animals were kept for the year to reintroduce manured soil for crop planting the next year. Such Common Field practices continued in a few parts of Essex such as Strethall into the 19th century. Many of our minor country roads owe their twisting course to the headlands of the Furlongs on which the ploughman turned his horses or oxen and plough at the end of each furrow. These unploughed headlands later became lanes and roads.

The word 'field' originally implied Open Land, not a hedged cultivated field at present understood. This land had at some period been cleared of trees; the early meaning of Ley meant woodland cleared of trees or scrub and later came to mean pasture. The early meaning of 'field' is contained in Hatfield, 'heathy open country'. Furlong names (and hence the present Field-names) often consist of two parts, the first a Reference Name which often is derived from a nearby feature such as a wood or stream or usual crop or soil characteristics. The second part is a

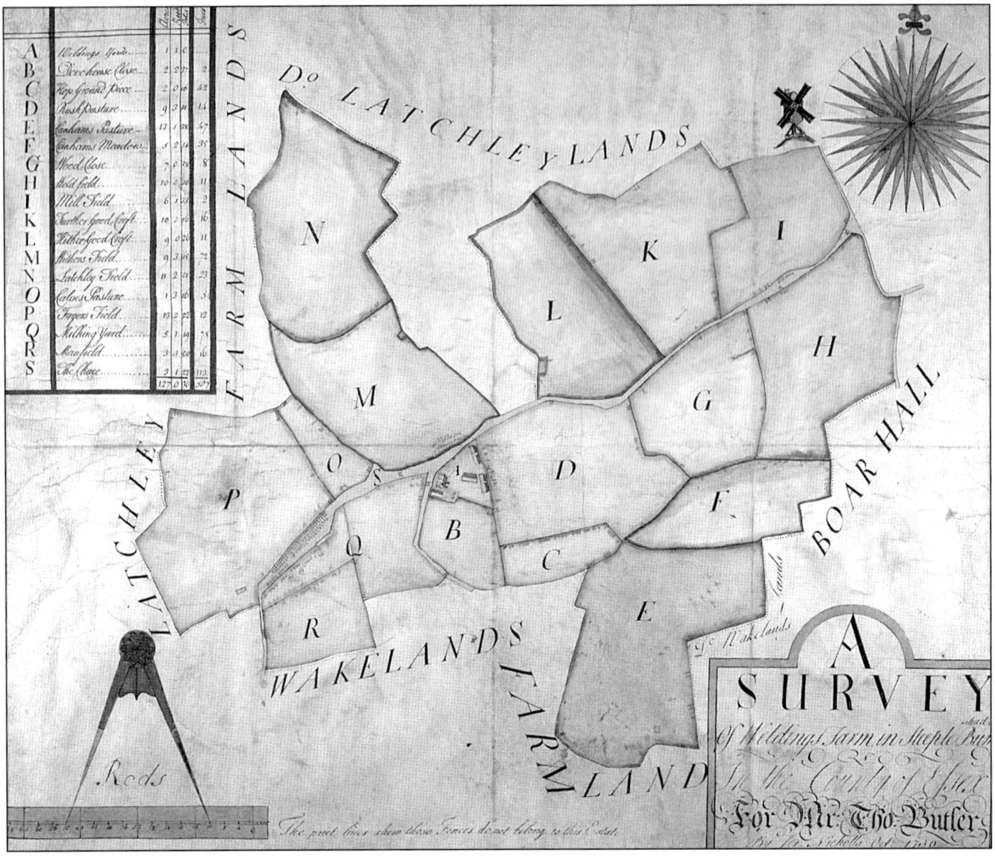

10. *Map of Weldings Farm, Steeple Bumpstead, 1737, with field-names.*

Qualifying Name such as Close, Down, Holm or Shaw, respectively, fenced or hedged land, low hilly open land, water-side meadowland or copse. Croft usually implied small enclosed land close to a dwelling. How land was used is indicated by maed, meadow pasture, aecer, ploughed, and Coppice. There are a large number of such Qualifying Names which, often taken for granted, can add much to the understanding of the old manor or parish system and geography.

Several attempts at Classification of Field-names have been made: Fertility names Small Gains (West Hanningfield), Potash field (Upminster) and Patience (Shelley), names indicating far-distance from the farm settlement America (Hatfield Broad Oak), Bombay Lands (Cressing), and Botany Bay field (in Great Leighs) etc. and the reader will devise their own understanding of the various types of Field-names as they are studied. Newly-cultivated land is described in Le Braches (Hornchurch); Butt field (Writtle) may indicate where archery was practised, a field still with tree-stumps or one which is shortened by being 'butted' between two adjacent fields. Husbandry and past activity are indicated by Sheepcote (Southminster), and Plaistow Green (Halstead), the sports-field. Crops are indicated in Sainfoin field (Wicken Bonhunt), Bannerley (bean-clearing) (in Cressing) and Ozier Bed (withies) (in Sturmer).

Industrial names indicate sites of previous activity, sometimes on a small scale. Pottekyns (Feering), Tylors Croft (Felsted), Dyers Ley (Writtle), Kiln field, Colliers Croft and, in Loughton, Coldbakers (charcoal-burning)) may still show evidence on the ground in the form of pottery wasters, brick and tile, or an uneven land surface concealing buried archaeology. Questers (Great Sampford), Tainter Pasture (Sturmer), Walk Field (Willingale Doe; walking or fulling) and Sawpit Field (in Stanway) were probably sites of quernstones, the cloth and wood-working industries.

Quarrying for agricultural purposes as well as for mineral and gravel extraction is recognised in Marl Pit Field (Stow Maries, marl, a mixture of clay and lime), Ochre Mead (Chignall St. James, an iron-rich soil used for marking sheep) and Copperas Ground (Brightlingsea, used for treating sheep-scab and in the dyeing industry). Crab Tree Delf and Ashen Delf (both in Stifford) indicate the location by notable trees of quarrying which continues still in the parish.

Just as with major Place-names, Field-names undergo evolution and, being often orally transmitted until documented much later than major Place-names, are prone to the effects of dialect and adaption to sound like but in written form distant from their original. Thus Tiltland in Copford has evolved from the 15th-century Tedelond, from Old English tyned, fenced. Pale Gate field (in Writtle) indicates not its colour but derives from Norman French pal, stakes enclosing land. Another name for enclosed land is Park from Norman French (e.g. Hog Park in Cressing) which denotes land enclosed by a ditch, bank or fence to keep in the deer (in distinction to Forest which was unenclosed). The name Tye is a common one in Essex. Deriving from Old English teag, small enclosure, it has acquired the meaning of 'outlying common' in this county. Around tyes often developed small hamlets whose tenants would use the common land for pasture.

Because Essex is abundantly supplied by rivers and streams it is not surprising that 'water names' figure predominantly. To allow flooding of meadows early in spring for early grass, many marshes, stream- and river-side fields were equipped with hatches or water-gates, as Marsh Hatch at Sickle Corner in Dagenham. The word needs to be distinguished from the hatch-gate to woodland (as Belzers Hatch field in Childerditch). Holme in Essex usually denotes a water-side or water-surrounded location, therefore land suitable for grazing. Since it occurs as an Old English word it cannot definitively be taken to indicate Viking influence, though in the Tendring peninsula or along the estuaries and coast it might derive from Old Norse holmr. The element hamm can also indicate land wholly or partly enclosed by water, for example by a river bend (for example possibly Short Ham and Long Ham in Panfield by Pods Brook), but its easy confusion with ham, settlement (already discussed) makes it often difficult to intrepret.

Marsh and Fen names are found near the several coastal estuaries. Saltings refer to the para-tidal and estuarine lands subjected to inundation by salt-water, and useful only on occasion for sheep-pasture. From at least the Iron Age through to the Roman period salt production was a major industry, by evaporating salt water. Carr implies the presence of alder trees, usually in wet areas, for example Alder Calf field in Cressing near River Brain.

Fords and bridges relate to local conditions; thus Rice Bridge in Little Clacton refers not to the Chinese staple but to Old English 'hris', brushwood. Oyster Bridge in Great Bardfield, Deresbridge in Great Leighs and Iron Bridge in Shalford deceive until their origin from 'eowestre', sheepfold, 'deor', animal, and 'eowu', ewe, is understood. Bridges over small rivers and streams probably postdate earlier fords, and the first documentation of a 'bridge' name at a previous 'ford' site, for example Bollifforde Bridge recorded in 1336 in Thaxted, marks the presence of a bridge no later than that date where at some prior date there had been a ford. 13th-century Trostelford in Inworth (Threshelfords Farm) is a happy reminder of the sound of the song-thrush by this crossing.

Local conditions are recorded in Fulebrige in Maldon (dirty), Horstedafort (now Stebbingford Bridge) (ford at the miry place), Claneforde at Wendens Ambo (cleared of rushes) and Caucebrigg in Aveley where the marsh by the Mar Dyke was approached across a causeway.

Perhaps on a more relaxed note, many field-names record where games were played and pastimes pursued. Dancing occurred on Dancers Close, Heydon, wrestling on Wrestlers in Thaxted and bull baiting on Bull Baiters in Great Leighs. Cock-fighting took place at Cockspits in Dedham. In the 13th century, every young person was required to practise archery in their parish; Archery Mead was probably where it took place in Harlow, and possibly in Butts Meadow in Downham. Festivities took place on Revels Field in Bobbingworth, quoits on Quoit Field in Toppesfield, unspecified games on Plaisters in Stanford Rivers and perhaps duelling on Draw Swords in Great Bentley. Football was played on Football Field (Nazeing and Bradwell-on-Sea). Some 'Camp field' names allude to an early form of football later to be banned as liable to breach the peace! In interpreting some of such names care needs to be taken to exclude a personal name.

Field-names provide clues to lost sites which prompt archaeological investigation. In Good Easter and Mashbury, 'the stronghold of Mæcca', Oldberry Field provides a focus for the site of the Saxon fortified enclosure (Old English burh, fort). At Rivenhall End, Lowes Field (hlaw, mound) contained evidence of a former Neolithic oval ditched burial tumulus.

Finally, as examples, Christian and pre-Christian customs have had influence on place- and field-names. Annual festivals (of 1st August of making the mass-loaf from the harvest) have left Lammas Land (in Chingford) and Maypole Field (Leaden Roding). The produce from Lampelande was given to provide lights in the church at Ridgewell, and from Bell Ropes for the church at Great Warley.

The research of Field-names offers opportunities for local study by individuals and local groups who are in a position to apply first-hand knowledge and field-study.

The Estates

Barking Monastery

Since some of the earliest documentation of place-names that has come down to the present day derives from early Anglo-Saxon charters, the brief consideration of a few of the charters of the Saxon Essex religious foundations is instructive. The seventh-century (*c*.675-8) charter of the endowment on Barking monastery survives (now in the British Library, Cotton Aug. II, 29; CS 81; Sawyer 1171). The single membrane of vellum, since 1938 between glass sheets but originally rolled into a scroll, later folded, written in Latin, measures 417x240mm. It is probably the earliest of all English royal diplomas to have survived in the original.

Bede, writing *c*.731, records the foundation by Erkenwald, probably a member of the royal household, of Barking during the reign of Sebbi, king of the East Saxons c.663-694. The traditional date c.666 is therefore likely. Soon after the foundation, Erkenwald was appointed fourth Bishop of London by Archbishop Theodore of Canterbury. His diocese included Surrey (which had formerly been part of the kingdom of the East Saxons) and that of the East Saxon kingdom which included Essex, Middlesex and part of Hertfordshire. His sister Æthelberga was abbess at the place called Berecingum (Barking). It was a monastic school having, in addition to nuns, also priests. As a foundation endowment for its maintenance, King Suidfrid granted 40 cassatae of land in Berecingas and Beddanhaam, the settlement of a person called Bedda. The boundary between Barking and Hamme was a stream (flowing east to the river Roding) which is called in the diploma for (East and West) Hamme, dated *c*.956, baede wyllan, Bedda's stream. The boundary between East Ham and Barking still follows this Back River to the west of the North Circular Road.

About 678, Hodilredus, a member of the East Saxon royal household, granted 75 (amended by a later scribe to 40) manentes (hides) at Ricingahaam, Buddinhaam, Dæccanhaam, Angenlabeshaam and a place in open country in woodland called Uuidmundesfelt. Of these Dagenham (part of Barking in Domesday Book, 1086) can be identified. Uuidmundesfelt is Wyfields, now a manor incorporated into (Great) Ilford parish. It included land between Cranbrook Road and Ley Street with the south-west part of Valentines Park (Lockwood, 2006). Buddinhaam may perhaps relate to Bedda (above), but the etymology is unsatisfactory. Ricingahaam and Angenlabeshaam await identification.

This charter had the boundaries of the land added by a later scribe c.820. These are:

ab oriente writola burna: on the east the chattering stream. (This is the Beam river which now enters the Thames at TQ499815.)

ab aquilone Centinces triow et hanc Hemstede: on the north Centinces tree and from here the homestead. (The present boundary meets the Beam (Rom) river at TQ517875. The tree perhaps in the region of Central Park TQ499869). Beorha's tree, on Becontree Heath and Five Elms were not far away.

ab australe flumen Tamisa: on the south the river Thames.

The western bounds are not given, probably because this land holding abutted the original holding of Bercingas and Beddanhaam. The boundary between Dagenham and Barking now runs northwards 300m (TQ490827) west of Dagenham Breach.

In 687, Bishop Erkenwald confirmed the privileges and grants already made to Barking monastery. This confirmation survives in a copy of the 16th century (BL Cotton Vesp. A ix, fo.112^{rv}: copy s.xvi^{1}; CS 87; S 1246). It specifies the original grant by King Suifrid of 40 cassati and the grant of Œdilred (alias Hodilredus), 75 hides of Ricingahaam, Bydinhaam, Daecanhaam, Angenlabeshaam and Uuidmundes feld, and ten hides called Celta. Celta is of particular interest for it may form the basis of 'Childerditch' and be the former name of the Mar Dyke. If so, these ten hides probably refer to land in Great Warley (3 hides), Bulphan (7 hides) and Stifford (40 acres) which were still in Barking's possession at the time of Domesday Book, 1086.

16th-century copies of two undated charters, but between 693 to 709, state that Suebred, king of the East Saxons, granted 30 hides in Nasyngum (Nazeing) of which part was called Enodmerese and ten called Ettunende obre to a person called 'Fymme', possibly a nun of the small community under the protection of Barking (Kelly Nos. 65a, 65b). In medieval times, Nazeing lay in the Deanery of Barking, though separated from the remainder by the Waltham Abbey jurisdiction (Bascombe, 1987). These two charters, discovered by Mr. H. Lockwood in 1969, have been linked with the two Saxon churches and cemetery (at Ordnance Survey reference TL386066) consisting mainly of females excavated in 1976, dated by radiocarbon to AD670+/-80 to 830+/-80. The community may have been destroyed by the Danes for it apparently did not survive into the tenth century. By 1062, Nazeing was granted to the community of the church of Waltham Holy Cross *(see below)*.

Only the southern boundary of the Nazeing estate is specified in the 16th-century copy of Suebred's charter which has survived, Stanhemstede, 'stone homestead', which is suggested to have been a homestead related to a Roman road leading to Fishers Green, TL378026. The 16th century copyist probably did not understand the Old English but Dr. Hart interprets Ettunende obre as at tun ende ofer, 'at the settlement at the end of the bank'. A possible candidate is the bank and ditch seen by Dr. Bascombe at TL410102, TL413097 and TL418086, south of Roydon village. If we use additional information contained in the charter of 1062 of Earl Harold's grant of Nazeing to found Waltham Abbey *(see below)*, the northern and eastern boundaries are provided. It provides an example of how investigation on the ground, archaeology and documentary evidence can combine to lead to identification of early estates.

Probably destroyed by the Viking invasions, Barking was refounded *c.*946 *(Fig. 11)*. From that year, three charters survive dealing with holdings at Hoccanleage, Hockley (Kelly No. 517b), land at Scopinglande, Shopland (ECE No. 12) and 4 hides at Tollesfuntum, probably a holding in the Tolleshunts and Tollesbury (Kelly 517a) of which 8 hides were in Barking's possession in 1086.

In 950, King Eadred granted to Barking four hides at Lippanwelle and four at Ciricdune (Kelly No. 552a). Half a hide (*c.*60 acres) at Limpwella, probably a copyist's spelling error, appears in Domesday Book held by Hugh from Odo, Bishop of Bayeux, which Edric a free man had held before 1066. A memorandum (ECE No. 106) in a Barking monastery gospel book *c.*1090 shows land of Gilbert, a man of the Bishop of Bayeux, in Stifford and is almost certainly this small holding; it maintains the association with the stream, welle. 30 acres and a meadow lay north of the brook, 12 acres south and 30 acres 'at the stone'. A Stone House is shown on the map of 1774 in adjacent West Thurrock at TQ565779. This suggests that Lippwanwelle lay to the southwest of Stifford and extended into Thurrock. Particularly to the south, the topography has been greatly altered by quarrying.

About Ciricdune there is more certainty. The Old English form is 'church hill' which is shown on the Ordnance Survey map c.1875 north and south of Chingford village (TQ374935). 250 yards south of the old Norman church of St. Peter's and St. Paul's was Cherrydown Farm, in the 1559 terrier Chear(e)downe, a corrupt form deriving from cirice, church. By Domesday, Barking had lost it and 6 hides in Chingford were held by the canons of St. Paul's in London and 5 hides by Ordgar, a thane of Robert Gernon.

11. The curfew tower at Barking, gateway into the abbey precincts.

St Paul's Monastery

Bede writes that Mellitus, whom Pope Gregory had sent in 601 to assist Augustine on his mission to Kent, was appointed first bishop to the East Saxon kingdom in 604. Its capital was London and on a hill above the Thames Æthelbert, king of Kent, built for Mellitus a church. No trace of this church nor of the later Saxon cathedral which was destroyed by fire in 1087 has yet been discovered. If the evidence has not been terraced away, it is likely to lie to the south of Wren's cathedral. Mellitus' mission was terminated when he was expelled by the pagan kings of the East Saxons in 616, though he later returned to Kent as the second Archbishop of Canterbury. Bishop Cedd, sent in 653 to return the East Saxons from idolatory, apparently did not take the London See, preferring to build new churches at Tilaburg (Tilbury) and Ythancaester (Bradwell, *Fig. 7*).

After his expulsion from the See of Wessex at Winchester by the West Saxon king, Wini was appointed Bishop of London (*c.*666 until *c.*675). Bede mentions St. Paul's church in London as the burial place of Sebbi, king of the East Saxons, who died c.694. Waldhere in 693 succeeded Erkenwald, the fourth bishop. In Bede's time (731), Ingwald was Bishop of London.

There exists, in a 12th-century copy, a list of estates held by of St. Paul's *c.*995-998 (ECE 28). For convenience it is known as the 'Ship List' as it records the number of men required from each estate to man a warship against the Viking incursions. The system presages that alluded to in Domesday Book for Dover, Kent and of the Cinque Ports of Kent and East Sussex from the 12th century. The value for place-name study is that St. Paul's 17 estates named in the Ship List can be related to those in Domesday Book in 1086. The Essex estates are: Ticc (?for Cicc, St. Osyth and The Sokens), Tillingha(m), Dunmæwan (Dunmow), Tollesfuntan (Tolleshunt), Næsingstoce (Navestock), Hinawicun (?Wickham St Paul's), Orseathun (Orsett), Ligeandune (Laindon), Seopinglande (Shopland), Westru(m) Orseathu(m) (West

Orsett), Bylcha(m) (Belchamp St. Paul's), Coppanforda (Copford), Holande (Holland), Suthmynster (Southminster), Claccingtune (Clacton), Hæthlege (Hadleigh) and Codanham (Codham Hall in Great Warley).

These estates, many lying along the coast or by the Thames, suggest that contact would have been maintained with them by water transport, and, more importantly, their produce from sheep and cattle farming could be brought to London. In 1086, 12 estates were held by the canons and 34 by the bishop (a number privately), plus houses and land in Colchester. Of the estates it held before *c.*1020, only Dunmow, Bradwell (part of the Tillingham estate), ?Asheldham, Shopland, Holland and Laver seem to be have been lost into the hands of Norman landlords by 1086.

Tillingeham is recorded as a foundation grant to Mellitus by King Æthelberht of Kent *c.*604-616 in a charter of dubious authenticity but probably containing authentic material. Southminster came into St. Paul's possession as part of King Suebred's grant of regio Deningei (Dengie) *c.*709 and Naesingstoc by King Edgar (for Æthelred) in 867. The Will of Theodred *c.*950, Bishop of London, granted to the canons Denesige, Tillingham, Dunmowe and Cic (CS 1008). King Eadwig gave ten 'cassati' of land at Orseathan in 957 and King Æthelstan *c.*971 confirmed their possession of land at Bylcham cum Wicham, Eduluesnæsa (The Naze and Sokens) and Runweolla (Runwell) (CS 737). The monks of Ely exchanged five hides at Holand with St. Paul's in *c.*984 for 4½ hides at Milton in Bedfordshire. After the date of the 'Ship List', Tidwoldintune (Heybridge) came into its possession about 1002 from the Will of ealdorman Brihtnoth's widow, Ælffæd, and further acquisitions, land at Barling and Chingford (in a dubious charter) and Weeley can be followed in the eight decades up to 1086. Thus can be traced the assimilation of estates, and their names, by the monastery over four and a half centuries. These estates provided wealth in particular from the prosperous activities of sheep farming and salt production by extracting salt from sea-water.

Waltham Holy Cross

About 1177, a secular canon recorded the early history of the Waltham church. He tells how Tovi, the staller (keeper of the king's stables) of Cnut found a holy stone cross at Montecute, Somerset, and had it taken to the church at Waltham *c.*1035. Tovi endowed the church with five manors in Hertfordshire and Essex (Alderton, Kelvedon, Loughton and land at Waltham). In 1060 Harold, due to become king and to be killed at the Battle of Hastings six years later, son of the powerful Earl Godwin of Wessex, established a college of secular canons granting 13 manors in Essex, Hertfordshire, Bedfordshire and Berkshire, the grants confirmed in 1062 by King Edward (the Confessor) who, with his queen, attended the founding ceremony. In Essex these grants were Passe felda (Paslow Hall in High Ongar), Welde (South Weald), Up mynster, Walhfare (Walter Hall in Boreham), Tippadene (Debden), Ælware tone (Alderton Hall in Loughton), Wudeforde (Woodford), and Næssingan (Nazeing). The estate of Costed (TQ585946), a manor in Walde of St Osyth Abbey, became Brentwood.

After the Battle at Hastings, the canons brought Harold's body to Waltham where he was buried. William I confiscated the church and granted it to the Norman bishop of Durham, Walcher, who built a new church (which has similarities with his other building, Durham cathedral). It was probably William II who restored the church to the canons, but in 1177, perhaps as a penance following the murder of Thomas à Becket at Canterbury, Henry II established a monastery of Augustinian regular canons which most of the secular canons of Harold's foundation refused to join. The monastery attracted significant grants. In 1177/8 it had gained Epping, land in Dunton, Lambourne, and churches in Loughton, Nazeing, Netteswell, South Weald, Woodford and Sewardstone in Waltham Holy Cross. By 1189 it had gained the manor of Stanstead Abbots, by 1207 churches at Alphamstone and Waltham, by 1230 the manor of East Horndon, by 1300 the manors of Takeley, Theydon Bois, land at

12. Waltham Abbey gateway, built c.1369, and the Cornmill Stream.

Boreham, Colchester with Lexden and Stanway, Roydon and Stanford le Hope. The monastery was dissolved in 1540 by Henry VIII, the richest in Essex and the 29th richest in England, valued at £307-0-1½d. *(Fig. 12).*

The foundation charter dated 1062 confirming the estates exists in copies of which that dated c.1220 (BL Harley 391, 33r-35v: copy s. xiii.) is probably the earliest (Sawyer 1036). The charter details, in Old English, the boundaries of the lands granted by locating points on a perambulation around the estate. These points can sometimes still be identified on the ground, making local exploration of the topography of 1050 years ago a fascinating exercise. A more restful investigation may be taken using the 19th-century Ordnance Survey maps, the Tithe maps and Awards (*c.*1840), (easily accessible through the Tithe Parish Place-name booklets, bound with accompanying maps, issued by the Essex Record Office), and Chapman and André's map (1777). Perforce many of the points are now indistinguishable and open to discussion.

The bounds of the estates are suggested:

Of Paslow Hall (Ordnance Survey Grid Reference TL580033 in High Ongar) *(Fig. 13)* perambulated in a clockwise direction:

1. *ærest of thære ealden hæcce æt freothene felde* – first, from the old hatch-gate at protected open land (probably the gate onto Norton Heath, at TL601041)

2. *into preosta hlype* – to the priest's deer-leap (?near TL590015, where the the Tithe Award of 1848 shows Deer Gate field, an entrance into Paslow Wood Common)

3. *into tham bece* – to the stream (which rises 200m north-west of Ashling's Farm, TL591013, and meets the parish boundary at TL586012)

4. *to standune* – (north-west along the parish boundary) towards Stondon Hall

5. *of standune to scealdeforda* – to the shallow ford (?now Hallesford, in 1875 Horseford, TL562023)

6. *of scealdeforda to coleboge welle* – to Colbegn's stream (?which runs westwards north of Chivers Hall to join the River Roding at TL561036)

of there welle eft into thære ealdan hæcce – then back to the old hatch-gate

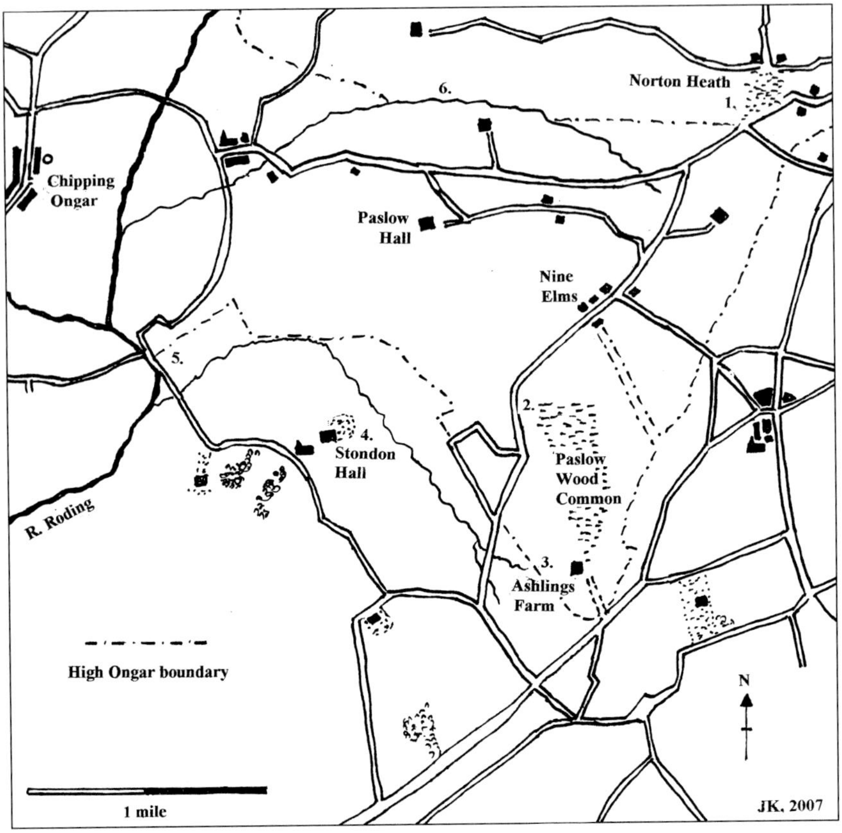

13. The Bounds of Paslow Hall, High Ongar, in 1062.

swa eft in freothen fælda – so then in the protected open land.

Of South Weald (including Brentwoood) proceeding in a clockwise direction:

erest of dellen north into there gemythe – first, from Weald Brook (crossing Hare Street, A12, at Putwell Bridge, TQ565923, north up the brook, following the Havering/Chafford Hundred boundary) to the confluence (of two streams, TQ944954) east into hafegeate – (along the Chafford/Ongar Hundred boundary) (north) east to (Great and Little) Howgate (fields, Tithe Award, c.1840, gate to common land at TQ556965)

of hafegeate east into tham wulfpitte – east (then northeast up Ongar Road) to the wolf-pit (near ?TQ569975)

of tham pytte suth into tham purce – south (east), across Kennel Lane, to Day's Farm, TQ586965, then east, crossing Brickhouse Wood, TQ595964, then south (Doddinghurst Road, and west (Hatch Road) to the '?paddock' (about TQ590958)

of tham purce suth to freobearnes hlype – south (along Beads Hall Lane, east of, then into the Pilgrims Hatch housing estates, (?Balmoral then King Georges roads) to Freobearn's deer-leap (probably the north end of Shenfield Common, TQ600935)

swa in to mannes lande – so (passing westwards by the Ursuline school fields to King's Road, TQ594931) to the Common

thanon eft into dellen – then back to Weald Brook (at Putwell Bridge).

Of Upminster in a clockwise direction:

ærest at tigel hyrste – first, at (Great) Tylers

(field) and Shaw (shown in the Tithe Award, *c.*1842, at TQ575914)

suth to thære mearcdice – south to the boundary ditch (along the Upminster parish boundary crossing the Hornchurch-Cranham road at TQ566867, then along Stubbers Lane)

of thære mearcdice west in ingce burne – westwards (from Running Water Brook, TQ573824, up Aveley road) to Ingrebourne river (TQ541853)

of there burne north into beccengare – (up the river) northwards into beacon triangle (Beacon Farm, TQ562902)

of beccengare north andlang thære weald strete into stangare – north along the wood street (?Hall Lane) into stone point (?where the Upminster/ South Weald parish boundary meets river Ingrebourne, TQ563916)

of stangare north into mannes lande – north into (Upminster) Common land

of mannes lande east into tigel hyrste – eastwards back to Tylers wood.

Of Walter Hall (*c.*TL738114 in Boreham) in an anticlockwise direction:

ærest of tham æssce to thære ældan hlype – first from the ash-tree to the old deer-leap (probably in the northeast of the estate (?TL755110)

of thære hlype to thære ealdan wude hæcce – to the old wood-gate (near TL736119 from woodland shown on Chapman & André's map of 1774)

of thære hæcce to thære ealdan strete – (along the parish boundary) to the old (Roman) road (from Chelmsford to Colchester at TL738091)

of thære strete to sand æcere – to Sandy Acres (shown on the Tithe Award *c.*1838, north of River Chelmer, TL748081)

of tham æcere to beadewan ea – to the Chelmer river. (The artificial Navigation at Stonham's Lock has altered the river's course here)

of thære ea to winibroca – (down the river) to Wine's Brook (which enters just west of the Boreham-Little Baddow roadbridge at TQ763085

of tham broce north eft to than æssce – northwards (up the stream) then back to the ash-tree.

Of Debden Hall (in Loughton) in a clockwise direction:

ærest in tippaburne – first on Tippa's stream (which runs out of Epping Forest at TQ436984 to Debden Green)

of thare burne up to tham hethe – upstream to the heath (around the Wake Arms cross, Theydon Bois-Waltham road junction with the Woodford-Epping road at TQ427995)

of tham hethe to thæcdene gemære – (turning north-east) to the Theydon(/Loughton parish) boundary (at TQ432996)

with æffan hecce swa into thære ea – (turning south-east at Jack's Hill along the boundary, leaving the forest) by the forest-gate (at Birch Wood (TQ443986) so to the river (Roding, TQ455966)

andlang thære ea west cymth in tippeburne – along (downstream) the river west till it comes to (its confluence with) Tippa's stream (at TQ452964, which, if correctly identified, constitutes the northern bound of the Alderton Hall estate).

By the 15th century the Tippaburne was called The Ree, no doubt deriving from Old English thære ea, the river. P.H.Reaney suggests that the Tippaburne was a tributary of The Ree which rose near Purles Lane, the confluence being north of Rookwood Avenue, though either interpretation is possible from the evidence.

Of Alderton Hall (TQ432962 in Loughton). The area is now so developed as to make recognition of most of the points difficult. Assuming a clockwise perambulation *(Fig. 14)*:

1. *ærest æt weard hæcce* – first at the watch hatch-gate (It is possible that the name of the Wake Arms crossing of the Waltham Abbey-Theydon Bois/Epping-Woodford roads has a very early origin. It appears on the 1st edition Ordnance Survey map of 1805 and 'Wake Valley' on later editions. A watch gate may have existed at this junction of ways through the Forest, TQ427994)

2. *of weard hæcce to ea crofte* – to the river croft. (A tributary of the Roding rises near Pyrles Lane at TL437973 to join Tippa's stream of the Debden

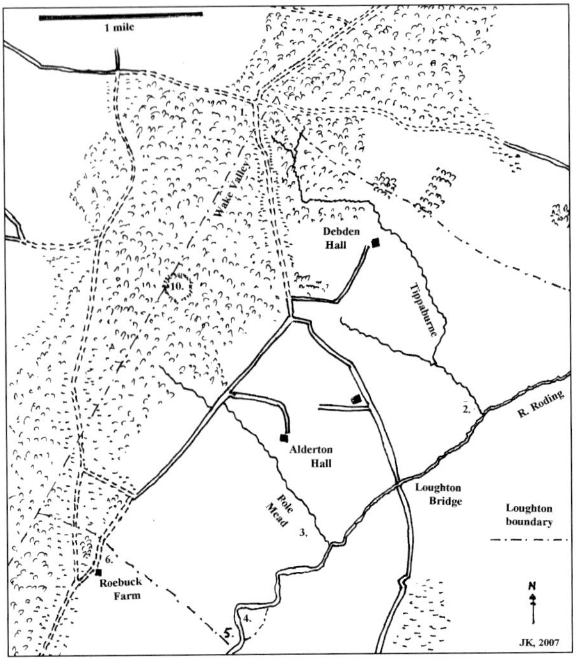

14. The Bounds of Alderton Hall, Loughton, 1062.

estate; it flows into the Roding at TQ452965). (See note about the Debden estate above)

3. *of æa crofte into beolle pole* – to the ?tree-trunk pool. (Following down-river, a long field, Pole Mead, TQ428953, ran southeast to the south of a stream (now on the south side of Roding Road) towards the river Roding. Near the river there may have been a pool)

4. *of tham pole into leofsiges mæd* – to Leofsige's meadow (It is tempting to equate this with Lushers Mead, TQ427945, in a meander now on the east bank of the river Roding but within Loughton parish as shown in the Survey *c.*1739 and 1820, Essex Record Office: TM 89/1)

5. *of leofsiges mæd into omer mæd* - to (omer, ? a bird of the family of) bunting meadow. The meadows are presumably near the River Roding (?Alderton Mead TQ426943 just north of Chigwell Bridge at Roding Lane).

6. *of omer mæd into ætherices hlype* – to Ætheric's deer-leap (probably following the Loughton parish boundary passing by Roebuck Farm TQ413946, back towards the forest)

7. *of thaere hlype into wulf hlype* – to the wolf-leap

8. *of wulf hlype into hæsfalde* – to the brushwood enclosure

9. *of hesfalde into stanweges hæcce* – to the stone way hatch-gate (This may indicate a hitherto unidentified Roman road through the Forest)

10. *of stanweges hæcce into sæteres byrig* – to the robbers' stronghold (?Loughton Camp, TQ418976).

Of Woodford proceeding anticlockwise:

ærest in angrices burne to ealdermannes hæcce – first, in Roding river (where the parish boundary crosses it at TQ419924, following the Becontree Hundred boundary across Lord's Bushes) to Alderman's Hatch on River Ching (TQ400942)

to thær cynges hæcce – (following the river southwest) to Chingford Hatch (TQ392930)

of thær cynges hæcce eft into angrices burne – then (following the parish boundary to Woodford Side, turning east at (TQ395904) back to the Roding river (TQ416897).

Of Nazeing perambulated clockwise from the north-west corner:

ærest of ceorlan hæcce – (probably) from the churl's hatch-gate (on River Lea TL390085, above Dobb's Weir west of Nether Hall) *and lang mearce to scealdan mere* – along the (Waltham Hundred) boundary to the shallow pond (at Didgemere Hall, TL407086)

scealdan mere to thære burnan – to the (Parndon) Brook (TL425075, east of Tylerscross)

of thære burnan to buter wylle – (continuing to follow the parish boundary) to the butter-stream (at Epping Long Green, TL425051)

of buter wylle to thuroldes gemæran – to Thurold's boundaries (yet to be identified, but Harold's Park Farm is at TL413045). The boundary follows Nazeing Long Green to TL375046, or Nazeing Brook which flows into River Lea at TL373062

of thuroldes gemæran eft andlang mearce to ceorlan hæcce – afterwards along the boundary back to the the churl's hatch-gate)

seo maed the thær to gebyrath lith ut beligean – the mead (?Nazeing Mead) that belongs to it out by the Lea.

Domesday Book shows that in 1086 the secular canons still held Passefelda, Welda, Vpmonstra, Walcfara, Tippedana, Lochintuna and Aluertuna, Wdefort and Nasinga, to which Epinga (Epping) had been added.

Ely Monastery and Nunnery

The double community on the Isle of Ely in the Fens was founded *c.* 673 by Æthelthryth, one of the daughters of Anna, king of the East Angles. It suffered under the Viking raids but was restored by Æthelwold, bishop of Winchester, under King Edgar, 959-975.

In 991, monks from Ely followed the English army at Maldon, and took the body of Brihtnoth, ealdorman of Essex, from the battlefield back to Ely. Ely held several estates in Essex and their acquisition can be traced by its charters. Between 996 and 1019, the lady Ætheliva gave Thacstede (Thaxted), in *c.*1010 lands in Rothinges (the Rodings) and a farm at Æthfelda (Hatfield Regis) were granted by Leofwine son of Æthulf, in 1004 twenty hides at Lyttlanbyrig (Littlebury) were given by King Æthelred (K711). Abbot Ælfsige purchased in 1008 an estate at Kadenho (Hadstock) which may have been the site of Cnut's minster, and Strethle, Stretley Green (K725). The monastery benefited under King Edward the Confessor who was brought up by the monks there.

The monastery suffered after the Norman invasion of 1066 for its tacit support of Hereward the Wake's resistance in the Fens. In spite of an Inquiry held by the king's Commissioners *c.*1075, in 1086 they still claimed Bedenesteda (Bensted in Sandon) which Hugh de Montfort took, a hide in Terling which Jocelyn Lorimer took, Phenbruge (South Fambridge) which Reginald Gunner took and Haneghefelda (South Hanningfield) which Thorold of Rochester took. High Roding and a hide in Rodinges (Leaden Roding) were taken by William de Warenne, Ratenduna (Rettendon) and a manor in Brochesheuot (Broxted) were taken by Eudo the king's Steward. Asgar the king's Staller (of the royal stables) took Estra (Pleshey then part of Easter) in 1066 and the Abbot claimed Morrell Roding which Geoffrey de Mandeville took, and Amb'dana (Amberden Hall in Loughton) which Ranulf Peverel took.

The Bounds of the land granted by King Æthelred to Ely monastery of Littlebury and Stretley Green are contained in Old English in the charter dated 1004 (Hesse, 1995). Proceeding clockwise, they begin at the river Cam at TL503427:

of tham forde the is Æthelwardes Ceasterforda gemera – from the ford which is Æthelward's boundary of Chesterford

and Ikelingtune gemera and Lytlabyrig gemera – and the boundaries of Icklington and Littlebury

swa andlang Styrice – then along [the river] Cam

into tham smalanforda – as far as the narrow ford (probably at Kelk's Bridge, TL521369)

of tham smalan forda to Streleage mere –from the narrow ford to Stretley mere (TL483383)

to crocesthorne – to croces thornbush (?TL472391)

of crocesthorne andlang sceacerdene – along robber valley (up to the edge of Free Wood)

into there ealdan delle – as far as the old pit

of there delle into tham hæcce – as far as the fence (?at Strethall Wood, TL480396)

into tham wulfpytte – as far as the wolf-pit

of tham pytte to Botolfes heale leage suthwearde – to the southern part of Botolf's corner wood

of there leage into Wullafes leage – as far as Wullaf's wood

of Wullaf's leage andlang into Leofsies pytte – as far as Leofsie's pit

of Leofsies pytte swa ut tham felda – out into open country (north Strethall was still Open Field in 1881)

into tham ealdan gehæge – as far as the old enclosure

of tham gehæge into Spelbeorhge – as far as Spelbeorg (perhaps the barrow next to Valance Farm, TL482418)

of Spelbeorhge into Styrice on – as far as the Cam at the

Icelingtune gemeræ and Ceasterforda gemeræ – boundaries of Icklington, Chesterford

and Lytlanbyrig gemeræ – and Littlebury.

Christ Church, Canterbury

Of the two early monasteries in Canterbury, Christ Church and St. Augustine's, both were founded by St. Augustine, Pope Gregory's missionary, who arrived in Britain in 597. In the Middle Ages Christ Church was known as Holy Trinity, now Canterbury cathedral, beneath which excavations have revealed the earlier Saxon churches.

King Cnut was a benefactor and, according to his Writ dated *c.*1020, personally visited the church. Christ Church acquired estates in Essex. The earliest grant of which we have details was in 823 when Leofstan granted Southchurch to the monks. Other early acquisitions were, by purchase by the archbishop from Ingelram, Vange in 963, and, from ealdorman Brihtnoth in 991 on his death at Maldon, Lawling in Latchingdon. With the consent of King Harthacnut, the priest Wulfstan granted Thurrock *c.*1042.

By 1086 the monks held land in Little Coggeshall, Bocking, Stisted, Lawling, Latchingdon, St. Lawrence, Milton Hall in Prittlewell, Bocking Hall in West Mersea, Southchurch and Little Stambridge. Wimbish which had been granted c.1043 by Thurstan, great grandson of Brihtnoth, was taken by Ralph Baynard, the Norman sheriff of Essex, before 1086. Count Eustace of Boulogne, brother-in-law of King Edward the Confessor, took Laver, St. Osyth, one hide of Ulham's Farm in Lawling, and nearly four hides in Coggeshall.

The warrior bishop Odo of Bayeux, half-brother of William the Conqueror, took Vange and Thurrock. Odo's trial at Penenden Heath, near Maidstone, *c.*1072, his forfeiture and imprisonment reportedly for trying to purchase recruits to assist him to seize the Papacy, resulted in Stisted, which had been granted to Christ Church between 1042 and 1066, and Little Stambridge being restored to the monastery, though Odo still was holding land in Great Stambridge in 1086.

King Edgar's grant of 7 hides at Vange to thegn Ingeram, dated 963, later the same year given to Christ Church, gives the bounds, in Old English, of Vange. They are suggested to be, proceeding clockwise, starting at

rum fleotes muthan, the mouth of the wide inlet, TQ737852,

andlang ælfwenne mearce, along Ælfwenn's boundary to (TQ710865)

tha stræte (Dry Street), (north) east along to

alfwoldes mearce, then north along the boundary

oth thone herpath, as far as the army road (Clay Hill Road?), east along to the bridge (a tributuary of the R. Crouch), north

andlang stræte (Timberlog Lane?) to *duddes hyrnan*, Duddes corner (TQ725894), and east to *cyneferthes mearce* (TQ734893), along *byrhtferdes mearce suth innon thone fleot*, Pitsea Creek, to the east of the *aesc hus*, ash house, along *fleot*, Vange Creek, to *rum fleot*, (the beginning point) and *tettincgleage*, wood.

The Battle of Assandun

"But these are deeds which should not pass away,
And names which must not wither".
George Byron, 1788-1824.

On 18th October 1016, the outcome of a decisive battle between the Viking forces of Cnut and the English army under King Edmund Ironside determined that the English throne should fall to a Dane, the first time England had not been ruled by Anglo-Saxons for over 400 years. The Anglo-Saxon Chronicle records that the battle was fought as the Vikings withdrew from Kent to Sheppey, thence into Essex on their way to Mercia (the Midlands). Edmund caught up with them in Essex at a hill called Assandun, where was fought a fierce battle. Because Ealdorman Eadric of Herefordshire and south Shropshire betrayed his king by taking his men from the battlefield, the flower of the English nation including monks from Ramsey and Ely monasteries who were at the battle were slain and Cnut was victorious. Within a few months after Edmund's death, Cnut was king of the whole country.

The two places which could qualify to have been Assandun are Ashingdon near the coast between the rivers Thames and Crouch, or Ashdon in the north of the county between the rivers Stour and Granta (Cam). If the Viking force had taken ship from Sheppey to the river Crouch, Ashingdon might be referred to. To have left their ships in the river Crouch on their way to Mercia so involving a long land journey across Essex seems less likely for a people more used to seamanship than land-trekking. If they took ship from Sheppey to the River Orwell junction with the Stour at Harwich (the here wic, '[Viking] army port') thence up the Stour then westwards overland to the Cam, Edmund Ironside may have confronted them at Ashdon.

The Old English place-names contained in early documentation (independant of the battle site) are:

Ashingdon: Nesenduna, 1086; Assendon, 1198; Essendon, 1203. The prefixing n may derive from atten esen dun, 'at the hill of ash-trees' or 'at the hill of the asses', or at tham Assan dun, 'at Assa's hill', a personal name otherwise unknown in England.

Ashdon: Estchentune, Æestchendune, *c.*1036; Æsredune, *c.*1043; Ascenduna, 1086; Essenduna, *c.*1100, 'ash-tree hill'. Rackham (1986) observes that the ash-tree favours limestone soil (as is found close by at Hadstock, rather than the clay and alluvium at Ashingdon). A hill-spur above the Bourne-Granta stream attains 354 ft (112m) just into Hadstock parish east of the common on a track which connected Hadstock church with Ashdon south of Hadstock Wood. But the place-name evidence does not definitively support one site or the other.

The Anglo-Saxon Chronicle, under the entry for the year 1020, also says that after the victory Cnut had built out of stone and lime a mynster church at Assandun for the souls of the men slain there. Neither the churches at Ashdon nor Ashingdon have standing features which can be dated to earlier than the 14th century. Warwick Rodwell points out that at the small church at Ashingdon *(Fig. 15)* the length-width proportions of the nave suggest an original two-cell church characteristic of the 11th century. The somewhat larger church at Ashdon has a nave (excluding the 15th-century arcades and aisles) which suggests an otherwise unremarkable origin in the 11th century. There is however only two miles to the north a large cruciform church at Hadstock characteristic of

15. St Andrew's 14th-century church and tower, Ashingdon.

10th and 11th century minster churches found elsewhere in the country, and this has been claimed as Cnut's minster sited on the place he had brought together his army before the battle at Ashdon. The Saxon minster is sited in a corner of a large churchyard as if inserted into a pre-existing enclosure. The estate was in the ownership of Ely monastery since Abbot Ælfsige purchased 2 hides here in 1008 from King Æthelred, and it continued in its ownership by Domesday in 1086. It was perhaps monks from Hadstock who were on the field of battle in 1016.

Two field-names in Ashdon and Hadstock parishes may be relevant to the discussion. In the old Norse poem in honour of Cnut, dated *c.*1050, the battle site is located north of Danes Wood. Although Danes Wood has not been located there are Danes Croft, Long and Short Dane (fields) in Hadstock parish. In Ashdon is a field by the Old Rectory called Old Church Field, at some distance from the present church. Dr. Rodwell (1993) suggests that Cnut's minster, if not at Hadstock, may be buried beneath Old Church Field.

The Forests of Essex

There were five or six 'Forests' in Essex. The word derives from Latin foris, 'outdoors', and medieval Latin forestis silva, 'the wood outside the enclosed park' *(Fig. 16)*. Essex forests included Kingswood (Kingeswuda), north of Colchester, including Cestrewald, (Pitchbury Wood), Danegris (the brushwood of the Dæningas tribe, covering parts of Woodham, Lawling and Dengie), Hatfield, Hainault (Henehout, the forest of the community of nuns of Barking monastery), Writtle (which included Barrow Wood and High Wood), Domsey (covering parts of Easthorpe, Feering and Marks Tey), and Blackley in Felsted. Epping Forest (including Wintry, Winterheie north of Epping town, the winter-enclosure) was part of the Forest of Waltham.

Writtle and Hatfield were Royal Forests until 1238. The king owned the grazing rights and the trees of Kingswood Forest. 'Forest', not

16. The Forests of Essex in the 13th century.

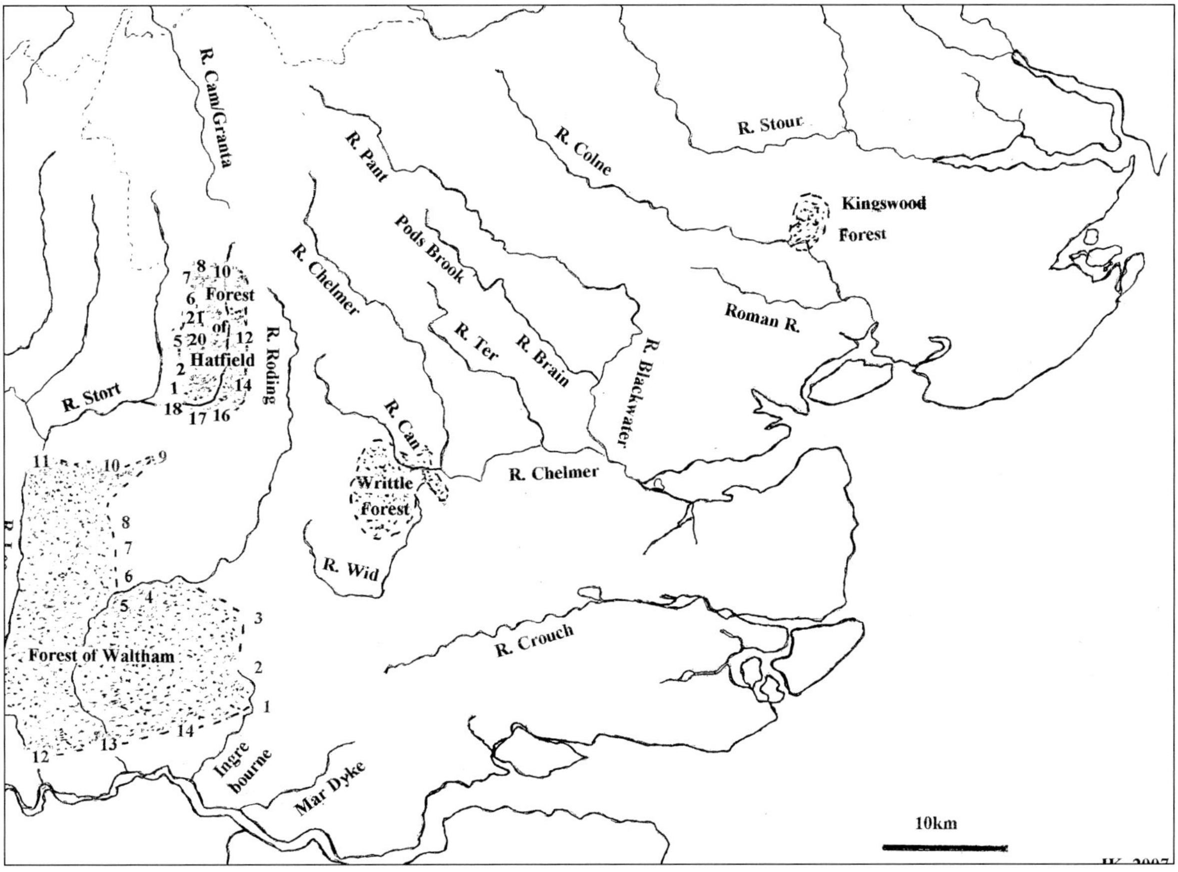

necessarily densely wooded and often heathland, was a legal concept designed for the keeping of deer and which retained certain rights to the owners of which, in some, ownership of the trees was one. The concept of Royal Forest was introduced to England by William I, and was developed by his son Rufus and Henry I (1100-1135). The Forest was administered by the Forest Court. There were gates across the roads which were designed to keep livestock contained. The position of gates is now remembered by many 'hatch' names (Pilgrims Hatch, Kelvedon Hatch, Chingford Hatch etc). Waltham Forest is referred to first *c.*1130 and survived pretty well intact from then until the 18th century when the encroachments and changes in definition of 'Forest' reduced it from some 60,000 acres to Epping Forest's present 6000 acres.

The extent of the '**Forest of Waltham**' was defined in the 13th century in two charters which deserve to be better known, dated before 1225 and February 1227. The boundaries were 'perambulated' by 12 knights, presumably on horseback. A conflation of the perambulations of the boundaries, starting at Putwell Bridge north of Romford and travelling anticlockwise, is:

"From **Putwell Bridge**, northwest up the **brook** between Havering-atte-Bower and South Weald to a **forge** on the edge of Navestock Wood, on to the River Roding, southwest to **Passingford Bridge** and **Aybridge**, then northwest along the **road** to **William de Bois's house** and his **wood** at Theydon Bois, to **Gregory Child's house**, straight on to the abbot of Waltham's **tiled house** by Holmencruke (at the cross south of the head of the North Weald boundary), then along the boundaries between the **Hundreds** of Ongar and Waltham, then Harlow and Waltham, to the **River Lea**, down to **Stratford Bridge**, then along the road to **Ilford**, **Romford**, and back to Putwell Bridge".

The points are :

1. **Putwell Bridge**, where the A12 crosses Ingrebourne River just southwest of the A12/M25 junction, TQ564923.
2. The **Brook** is Weald Brook (TQ5594), the name by which Ingrebourne river is known north of Putwell Bridge. It rises near Horseman's Side, TQ545960.
3. A **forge** by Navestock Wood, near Navestock Heath (TQ5397).
4. **Passingford Bridge** crosses the River Roding at TQ504975. There has been a ford here since before 1224. A bridge is first recorded in 1297 *(Fig. 17).*
5. The **Abridge** crosses the River Roding at TQ466970. Formerly Æffa's bridge; this Saxon name probably dates from before 1062, for æffan hecce is documented in a charter of that date.
6. The **road** is the Abridge Road, B172, to Theydon Bois.
7. **William de Bois house**, possibly Theydon Hall. The focus of Theydon Bois has shifted; the Hall and old church were at TQ463978, but the present church, *c.*1850, is at TQ448993. The Forest Gate is pinpointed by Chapman and André's map at TQ456990. Thrifts Hall Farm (TQ457987) was wood in 1777 and may be William de Bois's wood. The forest boundary then seems to run by the golf course to
8. **Gregory Child's house**, probably Gregories Farm in Theydon Bois (TL453005).
9. **Tiled house**: Ransford (1989) suggests Tyler's Green which is a cross-roads beyond North Weald Bassett, TL507051. The North Weald boundary follows Cripsey Brook ¾ mile to the north.
10. The **Ongar-Waltham Hundred** boundary runs south-west between the parishes of North Weald and Latton to TL436057 north of Epping Green.
11. The **Waltham-Harlow Hundred** boundary follows the parish boundaries south of Latton, Netteswell, Great Parndon and Roydon. The Hundred boundary reaches the **River Lea** at TL383080 southwest of Nether Hall. The course of the River Lea has been altered by the construction of the Navigation, but the old course is traceable from Chapman and André's map of 1777.
12. The main road from London crosses the old course of the River Lea at TQ386840 at

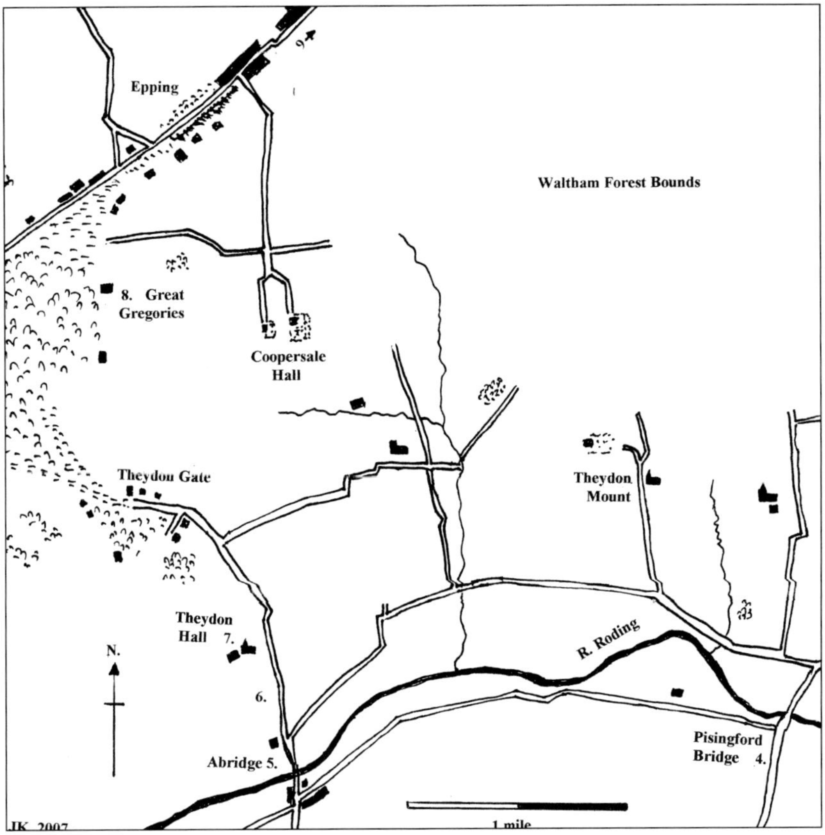

17. The 13th-century Bounds of Waltham Forest.

Stratford Bridge. The boundary follows the Old Romford road, A118, through

13. **Ilford** where it crosses the River Roding (TQ434863), to Seven Kings then

14. **Romford** High Street, Main Road to Gallows Corner (TQ534905) then along the A12, formerly Hare Street, back to **Putwell Bridge**.

The '**Forest of Hatfield**', still largely intact, is first specifically documented in 1225, as belonging to Henry III. The 'legal' extent of the Forest extended far beyond the physical forest and included probably most if not all the parish of Hatfield Regis. The boundary was perambulated in 1297/8 in a clockwise direction starting at:

"The end of **Shiringelane** between the lands once of Robert de **Shrowes** and Peter of Haselingfeud [in Cambs], then along the ditch separating Shiringe, Hatfeud and **Cowyks** manor as far as **Pokewoldeshach**, then along the ditch of Walter Attehach between Sabrightesworth [Herts], Hatfeild and **Little Hallingbery** to the pale of Little Hallingbury park, then along the boundary between Little Hallingbery and Hatfeilde to the ditch of **Robert Fitzwalter's park** in **Hallingebery bury**, then along the ditch through land called **Yemerlinoke** to **Thremhale** Priory, thence along the **King's Road** towards Dunmowe to **Martinesford bridge**, so continuing along the Road to **Henry Loveth's house**, then along the boundary between Hathfeud and Great Canfeud to the messuage formerly of Richard Moose, so continuing along the same boundary as far as a meadow called Holmad, then along the boundary to a pasture called **Littlehach**, and along the boundary between **Hathfeud**, **Heyrothinge** and **Whiterothinge**, to the bank and ditch of the wood of Wilfred Morel called Rowewod, thence along the boundary between

Hathfeud and Whiterothinge to **Walter Chalenar's messuage** on the boundary division of Whiterothinge, Matchinge and Hathfeud, along the Matchinge-Hathfeud boundary to **Hugo de Pervill's messuage**, continuing along the same boundary to the messuage of **Godfrey Attedonne** to **Gysebregge** and along the same boundary to **Wassingpol** bridge, thence back to **Shiringetye** at the end of Shiringslane. Including a wood called **Monkenewode**, next to **Walwode** in Litte Hallingbury".

The suggested points are:

1. the road to **Sheering** from Hatfield Heath at TL513142
2. **Shrubs**, TL516143
3. **Cowick** (cu wic, cow-farm, in Sheering), TL501151
4. the **hatch-gate** across the Hatfield Heath to Little Hallingbury road, TL512161
5. the edge of **Little Hallingbury park** at TL518165
6. i.e. **Great Hallingbury park** at TL527188
7. **Yemerlinoke**, a tree-name, TL530209
8. **Thremhall** Priory, TL531211
9. **Stane Street**, A120
10. **Martinesford bridge** over Pincey Brook at TL549211
11. house at TL565212
12. **Littlehach**, a gate (hatch) probably across Bexley Lane, TL574181
13. Hatfield, **High Roding** and **White Roding**, TL575156
14. **Rowe Wood** at TL556154
15. **Walter Chalenar's** messuage at TL545124
16. now **Parvill's**, TL535133, (but earlier in 1233 the home of Agnes de Pirevill)
17. now **Down Hall**, TL523131
18. **Gysebregge** (?Old English gyse, gushing water), **Wassingpol** bridge (?Middle English waschel pol, pool for doing washing) (Downhall? and Sheering bridges, TL512132)
19. **Sheering** Tye, TL506139
20. **Monk's Wood**, TL524176, (which belonged to the prior of Bermondsey).
21. **Wall Wood**, TL525185

Perambulation of the parish boundaries was also practised as the ceremony of 'Beating the Bounds' annually at Rogationtide (40 days after Easter) when the priest and his parishoners would walk the parish boundaries stopping for a prayer (some boundary fields acquired the name Amen field) and for children to 'beat' identification points with willow staves, so that parishioners would know the limits of their parish.

The '**Forest of Hainault**', a part of the Forest of Waltham, may derive its name from Old English higna holt, wood of the community (of nuns of Barking monastery). Havering Park (emparked or enclosed, often by ditch and bank, to keep in the deer) was part of the Forest but Havering-atte-Bower where there was a small palace was not. Havering Forest was granted to Queen Isabella, wife of Edward II (1307-27), for her dower. As in Writtle Forest, a chapel was established (by Henry VI in 1438). In 1851 the Crown abolished the legal restrictions on the Forest, permitted enclosure, and within a few years most of the woodland had been grubbed out. The woodland in 1774 is shown in Chapman and André's map (Fig. 18) extending from just east of Woodford Bridge (TQ432918) eastwards to the Havering-Redbridge boundary and Butcher's Farm (TQ505944), from Lambourne Road in the north to Billet Lane in the south. Woodland pasture extended along Barkingside (Tomswood Hill) to Little Heath (TQ468890) to Chase Cross (TQ511917). The Fairlop Oak was included within it. In all it then measured about 4900 acres of which only 300, mostly the eastern part, now survive.

Several gates to the Forest are shown, on the northwest Chigwell Hatch (TQ432925), on the south Abury Gate (Aldborough Hatch, TQ456893) and Rose Lane Gate (TQ480897, cp Rosehatch Avenue), on the southeast Marks Gate (TQ485905) and Collier-row Gate (TQ499912). Street names in Hainault remember the former Forest and its Court: Verderers Road, Regarder Road, Fowler Road, Rowbuck Road, Hind Close, Hart Crescent, Boar Close, Fox Burrow Road, Forest Road.

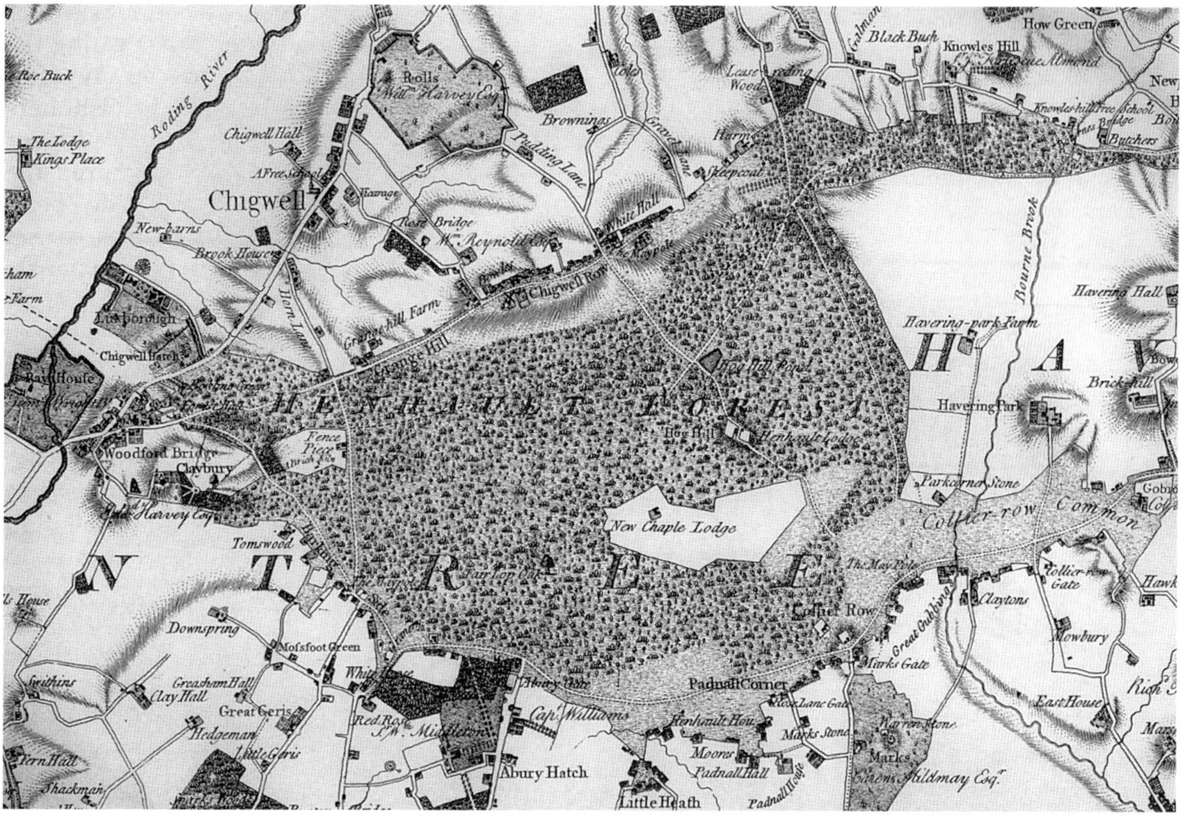

18. *Hainault Forest in 1774.*

'**Epping Forest**', also within the ancient Forest of Waltham, is first documented so named in 1662. In 1774 it measured about 6340 acres, some 700 acres more than now. The losses are most severe due to the development of Chingford, Woodford and Loughton. The Fairmead Lodge from which the forest administration was carried out, just south of High Beech (TQ407978), was pulled down *c.*1880 but the hunting lodge known as the 'Great Stonedeinge' from which the hunt could be watched, probably built for Henry VIII in 1543 and restored by Elizabeth I in 1589, was renewed. Epping Town was built on Epping Heath separating Wintry Forest to the north of it. Fairmead Bottom (Fairmead Road), the Forest between Snaresbrook and Whipps Cross, and the Lower Forest, now Wanstead Flats, are shown on the map of 1774 as wood-pasture or plain, for in the 13th century 'Fairmead' denoted grazing-land while 'High Wood' denoted woodland. New Road between Fairmead Road and Jacks Hill (TQ434995) was new at the end of the 18th century before which the London Road was the now minor road by High Beach.

Monkham's Lane led to Monkham's Farm, Monekenebukhurst, 1286, held by the monks of Statford Langthorne Abbey. Chingford Hatch was one of the gates to the Forest and Whitehall Road marks The White Hall on the Forest edge. Another Forest gate was the Theydon Gate where the railway now crosses the Abridge-Theydon Bois road. Clays Lane led to Golden Hill, the residence of Richard Lomax Clay, esq. in 1774.

The '**Forest of Writtle**' (anciently called Heywode) was a royal forest in the 12th century. Domesday Book records that it had woodland enough for 1500 pigs (i.e. a large area of wood) and that one of William I's followers appointed one of the resident pigmen to be Forester of the

King's Wood. The manor (recorded at 16 hides) was a large one and probably included Newland Hall now in Roxwell. It extended in an inverted horseshoe to include also the southern and western part of Hylands Park, the manor of Southwood as far south as Coldhall Lane being in the lordship of Chelmsford, property of the Bishop of London. Now the wooded areas lie mostly in the southwest of the parish which, in 1839, measured in all some 8410 acres (in 1875, 8786 acres including 650 acres of detached portions which were in 1888 transferred to the parish of Chignall). The king's law extended into adjacent Ingatestone and Fryerning for we read that in 1245 Richard de Munfichet was allowed to take 10 oaks from the king's forest of Ginges.

Personal names indicate forest husbandry, John le Forester, 1357, and probably Thomas Foster, 1659. There was a resident priest in King Harold's time before 1066 and a small hermitage in the 12th century, a subsidiary of St. John's Abbey Colchester, at Monks and Barrows (Farm) called Bedeman's Berg ('hill of the man-of-prayer', perhaps wth early confusion with bearu). There was a hunting lodge (now in the grounds of Writtle Agricultural College).

The Old English words for woodland include wudu (wood), leah (ley, wood or clearing in a wood), bearu (small wood), fyrhth (frith), grafa (coppiced wood, grove), hangra (hanger, wooded slope), holt (wood with a single type of tree or small wood, e.g. Hainault), hyrst (hurst, wooded hill), ryden (clearing), sceaga (shaw) and wald (weald). Middle English spring is a small wood. In Writtle, there are many place-names representing woodland, though most of the extant woods survive only at the periphery of the parish; these names may give an indication of the past extent of the physical forest. Starting at the north and moving clockwise, the following place-names suggest past and present physical woodland:

Moor Hall Wood (La More, 1248, Morehall, 1418), TL645063; Bushy Leys, TL649063; Upper, Lower Ley, TL650060; Bushy Lees, TL647060; Boy Ley, TL647058; Great, Little Boy Ley, TL645056. Long Ley, Middle Ley, TL622049; Wood Meadow, TL616046; Horsfrith (Horsfrith, 1213, 'muddy wood'), TL614044.

Wood field, TL626039. Parsons Spring, TL623027; High Wood, (Alto Bosco, 1274, Heywode, 1429), TL637025; Barrow Wood Birch (Forest of Berewe, 1323), TL630026; Birch Spring (cp. La Byrchhacce, 1291), TL625023; Deerslade Wood, TL636024; Coppice Wood, TL636027; Ellis Wood (John Elys, 1479), TL642028; Parkpond Wood, TL646027.

Hockley Shaw, TL651028; Writtle Park Wood, TL656030; Great, Little Edney Wood (Edenehay, 1254), TL657038; Upper, Lower Edney field, TL646044; Upper, Lower Furze field, TL656044; Writtle Park (parco de Writele, 1276), TL647035.

Lee Wood (Lalega, 1225), TL659048; Ley field, TL659047; Lee Spring, TL658047; Monks Ley, TL649045; Wood field, TL652049; Highland Grove, TL652051; Lady Grove (Ladygrove, 1523, of Mary, wife of Edward VI), TL653055. Wood field, TL664044; Spring field, TL665046; Brick Ley, TL664045.

Hedingham (Henningham) Wood (Hevynghams, 1412, from Heveningham, Suffolk), TL675051; Wood field, TL675052; Great Wood Field and Close, TL680037; Lightfoot Spring, TL681037; Hylands Park (west), TL681042.

College Wood (of New College, Oxford), TL690086; Great Wood field, TL690092; Great, Little Bushetts, TL688092; Broom Wood, TL688098. Eight Acre Ley, TL680085; Six Acre Ley, TL680086; Three Acre Ley, TL678087; Great Wood field, TL676085; Upper Ley, TL672085; Lower Ley, TL671085. Bushy Wood, TL687106; Bushey Lees, TL687105; Broom Wood, TL689097.

Extending across the parish boundary into Ingatestone and Fryerning we find:

Redindyke (Writeledich, 1297), TL648025; Wells Farm (le Welle redyn, 1502); Finche's Spring (John ffynche, 1450), TL645027; Box Wood, TL643017; Stoneymore Wood (graua voc. la more, 1291), TL636016.

Bell Grove, TL624014; College Wood (of Wadham College, Oxford), TL624018; Mapletree Lane (cp. Mapelereshach, 1297, a Roman road), TL623013; Fryerning Wood, TL615012; Portsmoorhall Wood (Possemor, 1291, Possa's moor), TL613008; Woodbarns, TL620008 and Furze Hall, TL623009.

'**The Forest of Colchester or King's Wood**': Like Writtle and Hatfield this was a royal Forest. It was granted by Henry I or Stephen to the burgesses of Colchester, and Henry VIII granted it to the town bailiffs. It included the intramural part of the town as well as land to the north and northeast. Its Bounds were perambulated in 1301 but are not easy to follow. They begin at:

"**Northbregg** by the king's highway to **Mylandecrouches** at the cross at Kynggeswodeheth, thence before the house of Richard Martyn to the hedge before the house of William Waryn of Colchester. Thence by the ditch of the heath to open space **Kynggeswodehach** between the vills of **Horkeleye** and **Kynggeswodeheth**. So straight by the ditch of Phillip at Hacche to the house of William le Herde of Horkeleye, so by the ditch between **Horkeleye** and **Kynngeswode** to the boundary between the vills of Horkeleye, **Bexsted** and **Kynggeswode**. Then by a ditch between Bexsted and Kynggeswode to the boundary between the wood of **Ralph de Bexted** and the wood of **John le Breton**. Then as far as the ditch of the **park of Lengham**. Then as far as the **Kynggeswodebregg** at the side of the park. So returning by the king's **highway** towards Colchester as far as **Estgate**, and all the **desmesnes** of the king and his castle within the town".

The medieval Liberty of Colchester incorporated King's Wood and assists in the identification of some of the boundary points which perambulated clockwise are suggested:

1. **Northbridge** (TL994256) over the River Colne to the north of North Hill, Colchester.
2. Along the **Colchester-Mile End road**.
3. **Mile End Cross** at the road junctions where the roads to Great Horkesley and Boxted Heath diverge (TL990281)

4,5,6,7. These points probably lie west taking the boundary towards **St. Botolph's Brook**, TL974274, then northwards to

8. The meeting of the Horkesley/Boxted/King's Wood boundaries is northwest of Cuckoo Farm (TL986293).
9. Turning eastwards along the **dyke** (possibly of Iron Age origin to **Salary Brook**). There were several woods south of Langham Park in the 18th and 19th centuries of which only **Kiln Wood** southeast of Langham Lodge has persisted in anything like its 19th century form (TM016294).
10. **King's Wood Bridge** over the western arm of Ardleigh reservoir (at TM022289)
11. Thence southwards along the Roman road passing Brickhouse Farm and Parsons Heath
12. into East Street (TM008253) and **East Gate** (TM005253). The road pattern south of Dilbridge has changed as a result of the railway. Formerly the Ipswich road joined directly into East Street.

It will be appreciated that many of the above place-names are preserved to the present day. Some indicate the topography at the time they acquired their name, others contain a record of a past owner or tenant. Clearance of wood for cultivation of crops or pasture is suggested by 'Ley' names, and occasionally a date can be ascribed to the assart (clearance), such as shown in a surviving charter of *c.*1225 of Redindyke in Ingatestone by John de Geyton, steward of Barking Abbey.

Look-outs and Beacons

When the quickest communication by letter or dispatch across the country was by horse, an interconnecting system of fire-beacons which could be lit giving warning of imminent attack provided a rapid alert to muster forces.

The Old English word used for a Look-out house was tot-ærn and for beacon, (ge)be(a)con. The Anglo-Saxon Chronicle in 1006 refers to the Viking army lighting here-beacen. In Essex the earliest references to a place containing tot are in the Will of ealdorman Ælfgar, dated *c.* 950 (CS 1012) and the Will (dated *c.*1000) of Ælfflæd, widow of ealdorman Brihtnoth who died at Maldon fighting the Danes (CS 1289). Ælfgar had left his estate at Totham to Brihtnoth. The tot referred to is probably at Beacon Hill in Totham which rises to 82 metres or perhaps Totham Hill a little to the east.

The foundation charter of the Waltham religious community, a gift of Earl Harold in 1062, includes the bounds of Upminster. In the northwest corner is a place called beccengare, 'beacon triangle of land', not far from Beacon Farm. 'Tot' and 'beacon' names are also found in documents containing field-names, and, while archaeological evidence for these ephemeral structures (which may have been no more than brushwood bonfires, though later 'beacon-trees' were fire-buckets filled with combustibles and pitch hoisted onto stout poles) may be slight or absent.

The beacon system to which look-out stations must have been linked clearly depended on its success on continuity of sight-lines from station to station. Hills and elevations often provided such stations and though some research into the probable sites of former beacons can be carried out from maps showing height-contours, there is little substitute for on-the-ground investigation. Trees and buildings in a county where hills are often low may now compromise former sight-lines so that reference to earlier maps which may show woodland may be useful even when contours are not depicted. It should not be forgotten that there is evidence that church towers provided look-out points and were also used for signalling. Ogilby depicted beacon structures on his road maps dated *c.*1675 (at Stanway Beacon End) and his maps of c.1678 at West Thurrock, Foulness, Langdon Hills and South Benfleet.

Place-name evidence is provided in Domesday Book for a possible beacon site (or from a personal name Becca), Bacheneia, later Beckeneye, at Ashingdon by the river Crouch marshes. Plumberow Mount in Hockley may have been used as a signal mound. These, within sight of Beacon Hill in Cold Norton and of another Beacon Hill at Canewdon, could have formed part of an integrated system. It is by study of the combination of topography, place-names and archaeology that systems of message transfer using signalling and beacons can be accumulated.

The Saxon moot or meeting-place for the Hundred of Becontree may have been near its geographical centre on Becontree Heath east of Valence House. An origin from beacon-treow is possible here, a reference to the pole or post at the top of which a fire was lit as a beacon. Another possible site is above Cobbins Brook in Epping referred to in 1386 as le Bekyn, later remembered in Bacons Field. This may have communicated with another at Bacon End in Great Canfield and Totegrove in Birchanger.

Thus by the tracing of place-names can be built up evidence for the systems of interconnecting beacon and look-out sites which have now little or no evidence remaining on the ground.

Dictionary of Essex Parish, Manor and River names

*"But still the heart doth need a language,
still doth the old instinct bring back the old names".*
Samuel Coleridge, 1772-1834.

A small handful of some of the earliest written references to Essex place-names is contained in the priest Bede's *Historia Ecclestiastica* written in Latin *c.*731. A few more appear 160 years later in the Anglo-Saxon Chronicle written in Old English under King Alfred's patronage *c.*891, with later additions.

Many place-names take origin from the name of an owner or tenant. Though it is intellectually unsatisfying, it has to be accepted, until further evidence is found, that little or nothing may be known about these people. A minority can be identified with the church (bishops, priests), the nobility (Saxon ealdormen or thegns), from their Wills, land grants or as witnesses to Charters, Deeds and other transactions. Occasionally such documents are rediscovered, such as two previously unknown Charters of Suebred, king of the East Saxons, dated *c.*700 relating to Nazeing, which were found in Hatfield House in 1969 as reused vellum in a 16th-century manuscript of Ilford Hospital.

Domesday Book compiled for William I in 1086 contains many of the earliest-known names. *Inquisitio Eliensis* (IE in the subsequent text) is a record resembling Domesday Book contained in several documents of the estates held and (with some justification) claimed by Ely Abbey. It is dated *c.*1090 with the copies surviving of the late 12th century. Many of Ely's lands were appropriated after 1066, particularly by Esgar the Staller, and the Abbey continued to claim them, sometimes with success, often without.

For people who ventured rarely beyond their own immediate area or settlement, when transport was mainly on foot so that a day's journey and back was no more than 20 miles, it is unsurprising that rivers had different names along their length. Some of present river-names can be dated no earlier than the 15th or 16th centuries or later, thus River Brain near Witham was 'aqua de Wyham' in the 13th century, and 'ripa de Nottele' near the Notleys in the 14th. Not until the 19th century is 'Brain river' documented for its length.

The Saxon land-taxation units were the 'Hundreds', of which latterly Essex had twenty (small units were sometimes called Half-hundreds). Originally they probably represented one hundred hides, a hide being a tax-unit of land nominally enough to support a family, in Essex perhaps about 120 acres, though the area depended on the fertility of the fields. Each Hundred held a monthly 'moot' at a convenient place near the centre; decisions were made and proclamations announced. The Domesday Book refers to Hundreds and to manors each held by a lord-in-chief, perhaps the king or a bishop or a religious establishment, from whom a lesser person held the tenancy.

The parishes we now recognise probably originated in the 11th and 12th centuries. Prior to that the small unit of landholding was the manor, and many of the names we now use for parishes originated before the 11th century as the names of manors. The following list gives the first recorded reference, the word element (often Old English) and the probable meaning. New discoveries of earlier records with different spellings have led to re-evaluation of these meanings and it is quite possible that further discoveries will lead to new interpretations; there will not be certainty about word meanings and debate will continue.

The documentary source (abbreviated, *see* Appendix) immediately follows the document date, for example, 1086,DB means the entry is found in Domesday Book. Where the word element is other than Old English, this is stated. People whose names appear in the chapter 'Who's Who' *(page 146)* are in **bold type** where further information is provided. The location of some French place-names is shown in *Fig. 6***.**

ABBERTON
Ralph of Marcy (La Manche, France) held *Edburgetuna* in 1086,DB: tun, Eadburh's settlement. *Fig. 6.*

ABBESS RODING
In 1237,Colch, *Roinges Abbatisse* was the property of the abbess of Barking monastery. *See Roding.*

ABRIDGE
Affebrigg, 1203,Cur: brycg, Æffa's bridge (over River Roding).

ALDHAM
Aldeham was held by **Odo**, the Bishop of Bayeux in 1086,DB: eald ham, Ealda's or old settlement.

ALPHAMSTONE
Bury St. Edmund's Abbey held *Alfelmestuna* in 1086,DB: tun, Ælfhelm's settlement.

ALRESFORD
*c.*1000, (CS 1289), **Ælfflæd**, widow of ealdorman **Brihtnoth**, willed *Ælesforda* to King **Æthelred**. Bishop of London and **Odo** held *E(i)lesforda* in 1086,DB: el ford, ford (through Alresford Creek) where eels are caught.

ALTHORNE
(Walter de) *Aletor*, 1166,Paper; *Aledhorn*, 1198,FF. In 1203,FF, Eustace fitz Terriel quitclaimed to William de Ponte fracto one caracute of land in *Alethorn* for 8 marks of silver: æled thorn, thorn-bush by a beacon or burned.

ARDLEIGH
Geoffey de Mandeville held *Erleia* in 1086,DB: eard or erth leah, wood/clearing with a dwelling or ploughed.

ARKESDEN
Eudo the King's Steward held *Archesdana* in 1086,DB: denu, Arcel's valley (of Wicken Water). *See also Chardwell.*

ASHDON
*c.*1036, **Lustwine**, grandson of ealdorman **Brihtnoth**, willed *Æstchendune* to Ely monastery (LE). His son **Thurstan** willed *Æsredune* to his wife Æthelgyth *c.*1045 (ECE 59). **Ralph Baynard** held *Ascenduna* in 1086,DB. *c.*1100, (Dugd), he granted the church at *Essenduna* to Lewes priory, Sussex: æscen dun, ash-tree hill (above River Bourn). *See also Ashingdon.*

ASHELDHAM
Hainctuna was held by **Swein** of Essex in 1086,DB: heah ing tun, high-place settlement. It lies above the marshes. But by *c.*1130,(Bodl), *Assildham*: (a woman) Æschild's settlement.

ASHEN
Richard son of Count Gilbert held *Asce* in 1086,DB. In 1090,(Dugd), Gilbert gave the church at *Essa* to Clare priory (Suffolk): æsc, (at the) ash-tree.

ASHINGDON
King **Edmund** and **Cnut** fought a battle at *Assandune* in 1016,ASC. There is disagreement whether this was Ashdon or Ashingdon; **Swein** held *Nesenduna* in 1086,DB: assa dun, Assa's or ass's hill. *(Fig. 15). See also Beckney.*

ASHLYNS (GAPPS) HAMLET, HIGH ONGAR
Astlens, 1456,ECP: Richard *Ascelyn* claimed a messuage, 120 acres of land, 9 acres of meadow and 5 acres of pasture in 1321,FF. *See also Ongar and Paslow Hall.*

AVELEY
John son of Waleran held *Auileia* in 1086,DB: leah, Ælfgyth's wood/clearing. *See Bumpstead.*

19. 14th-century church with Norman front door, Little Baddow.

20. Ballingdon Hall, since 1835 in Suffolk.

AYTHORPE RODING

Rohinga Willelmi fil'Ailtrop, *c*.1200,AddCh: Hrotha's peoples' settlement, Aitrop son of Hugh made donations of land here to Colne Priory in the 12th century. *See Roding.*

BADDOW, GREAT, LITTLE

c. 989,(CS 1306), **Ælfhelm**, a minister of King Edgar, confirmed in his Will his gift of *Beadewan* to his wife. **William** I gave (Great) *Becdowa* to Caen nunnery *c*.1080 (ECE 77) which it held in 1086; Roger God-save-the-Ladies held 1½ hides in (Great) and Count **Eustace** held (Little) *Baduuen* in 1086,DB: Celtic river name (for River Chelmer). *(Fig. 19).*

BALLINGDON

Peter of Valognes held *Belindune* in 1086,DB: bealg dun, round hill (above River Stour). *(Fig. 6, Fig. 20).*

BARDFIELD

Richard son of Count Gilbert held *Birdefelda* in 1086,DB: byrde feld, open country on banks (of river Pant). *(Fig. 21).*

BARKING

c.666, King **Suidfrid** granted *Berecingas et Beddanhaam* to found Barking monastery (ECE 2). The monastery held *Berchingas* in 1086,DB: ingas, settlement of Berica's people; ham, Bedda's homestead. *(Fig. 11).*

BARKINGSIDE

Thomas Grene, yeoman of Chigwell, willed 14 shillings owing in rent to William Batty of *Barkingside*, 1538,TEASxi,13.

BARLING

In 998, **Leofwine** willed *Bælingum* to **Wulfstan**, bishop of London (ECE 32). The canons of St. Paul's held *Berlinga* in 1086,DB: settlement of Bærla's people.

21. Little Bardfield Saxon tower and church.

22. Plan of Belchamp St. Paul manor house, church and mill, 1576.

BARNSTON
Bernestuna was held by **Geoffrey** de Mandeville in 1086,DB: tun, Beorn's settlement.

BASILDON
Swein of Essex held *Be(r)lesduna* in 1086,DB: dun, Beorhtel's hill (above Thames valley).

BEAM RIVER
*c.*678,(CS 81), *Writola burna* bounded Barking monastery land on the east: writol burna, chattering stream. *le Markediche*, 1154-89,HPD: mearc, boundary, dic, stream/ditch (between Havering and Becontree). *Pontem voc. Dagenham Beem*, 1299,WmC; beam, plank bridge. Lower part of Rom river, *qv*.

BEAUCHAMP RODING
Roynges Beuchamp, 1238,SD: John de Bello Campo, 1233, of the Bedfordshire family. (cp Belchamp Walter). Nicolas son of John de Bello Campo in 1321,FF, quitclaimed 1 messuage, and 1 caracute of land. *See Roding*.

BEAUMONT-CUM-MOZE
*c.*1000,(CS 1289), **Ælfflæd** willed *Fulanpettæ* to King **Æthelred**. **Aubrey** de Vere held *Fulpet* in 1086,DB: ful pytt, foul hollow. *Bealmont*, 1175,BL: Old French, beumont, beautiful hill. (Presumably the new hall was built on higher ground west of the marshland of Hamford Water.) **Geoffrey** de Mandeville held *Mosa* in 1086,DB. In 1087 **William I** confirmed a herring fishery at *Mosa* to be the property of Westminster Abbey: ?Scandinavian mosi, bog.

BECONTREE
Beuentreu, 1086,DB: treow, Beohha's tree or (ge)beacon treow, signal beacon-stand (which may have been on Becontree Heath east of Valence House).

BELCHAMP OTTON
Count **Eustace** held *Belcham* in 1086,DB. *Belcham Otes*, 1256,Ass: bylc ham, settlement with a roof-beam house; William filii (son of) **Otonis** held it in 1212; the family managed the royal mint.

BELCHAMP ST. PAUL'S
*c.*970,(CS 737), King **Æthelstan** confirmed the grant of *Bylcham* to St. Paul's which *c.*998 required 1 man for ship duty from *Bylcham* (ECE 28). The canons held *Belcham* in 1086,DB: bylc ham, settlement with a roof-beam house. *(Fig. 22)*.

BELCHAMP WALTER
*c.*1040,(K931), **Leofgifu** willed *Belhcham* to her Lady. **Aubrey** de Vere held *Belcamp* in 1086,DB: bylc ham, settlement with a roof-beam house. An ancestor of Walter de Teye, 1297,M, was William de Beauchamp. *(Fig. 23)*. *See Beauchamp Roding*.

23. *Belchamp Walter Hall, built 1720 (photo by courtesy of the owners).*

BENFLEET

In 894, the Anglo-Saxon Chronicle tells us that the Viking fleet was destroyed at *Beamfleote* by the English. **William I** granted 1 hide in (South) *Benfleota* to St. Martin le Grand, London in 1068, and the church at *Beamflete* to Westminster Abbey *c.*1080. Before 1066 **Harold** but in 1086,DB, **William I** held *Benflet.* St. Peter's Westminster and **Swein** held (South) *Benflet* in 1086,DB: beam fleot, creek with timbers (?a bridge probably near the present one at Benfleet Creek). *(Fig. 24). See Fanton Hall.*

BENTLEY

*c.*1040,(K931), **Leofgifu** willed (Great) *Benetleye* to Bishop Ælfweard, bishop of London. In 1086,DB, **Aubrey** de Vere held (Great) *Benetleia,* Count **Alan** held (Little) *Benetleia:* beonet leah, bent-grass clearing.

BERDEN

In 1087,WBD, **William I** confirmed the grant of *Berden* to Westminster Abbey. This charter may be a forgery. In 1086,DB, **Swein** held *Berdane*: bær denu, woodland-pasture valley for swine. *(Fig. 25).*

BERECHURCH/WEST DONYLAND

Berdechirche, 1216-72,Colch: bred cirice, (weather-)boarded church.

24. *The creek at South Benfleet.*

25. Berden Hall, a 16th century building. (Photo by courtesy of the owners)

BERNERS RODING

Reynges,Rothingg Berners, 1235,Ass: Hrotha's peoples' settlement, Hugh de Berners from Bernières in Calvados, was tenant in 1086,DB. *See Roding. (Fig. 6).*

BILLERICAY

Byllyrica, 1291,For: (Persian) balilah, a tree-fruit myrobalan used for dyeing, hence dye-house. *(Fig. 26).*

BIRCH, GREAT, LITTLE

(Great) *Bricia* was held by Count **Eustace** in 1086,DB, (Little) *Bricia* by **Robert** Gernon, *Parva Bricceia* by Hugh of St. Quentin (La Manche): bryce, recently broken land.

BIRCHANGER

King **William** held *Becangra* and **St. Valéry** *Bilichangra* in 1086,DB: bierce hangra, birch-covered slope. *(Fig. 6).*

BIRDBROOK

Ranulf brother of Ilger held *Bridebroc* in 1086,DB: bridd broc, birds' stream (a headwater of River Colne at Wash Bridge). *See Baythorn.*

BLACKMORE

In 1086,DB, **William I** held *Phinghería* (Fingrith): rith, stream of Fin's people (Wid Brook, see Writtle). *Blakemore,* 1213,Abbr: blæk mor, black marshland.

BLACKWATER RIVER

*Pantan stream, c.*1000?,Battle of Maldon poem; and *c.*1046 (ECE 63); *le Blak Water,* 1477,Ct. (Above Braintree it is the Pant, *q.v.*).

BOCKING

*c.*998,(Dugd), **Æthelric** and his wife Leofwyn granted *Boccinges* to Christ Church Canterbury. The monks held *Bochingas* in 1086,DB: settlement of Bocc's people. The former settlement was at Churchstreet.

26. Billericay High Street and 16th-century church tower.

BOREHAM
*c.*1040,(K931), **Leofgifu** willed *Borham* to her kinsman Ælfgar. Count **Eustace**, William de **Warenne** and **Swein** held *Borham* in 1086,DB: bor ham, settlement on a hill. *See Walter Hall.*

BORLEY
The Countess of Aumâle, half-sister to **William I**, held *Barlea* in 1086,DB: bar leah, boar clearing/ wood. *(Fig. 6).*

BOWERS GIFFORD
In 932 King **Æthelstan** granted 10 hides in *Bura* to Abbot Beorhtsige (?of Benfleet) (Kelly 418a). In 1065, King **Edward** confirmed the grant of *Bure* to Westminster Abbey. In 1086,DB, Westminster had 50 acres in *Bura*: bur, dwellings, William Giffard, 1243. The Gifford family was descended from the maternal line of **William I**. Sir John Gifford fought at Crécy in 1346.

BOXTED
Count **Eustace** held *Bocchesteda* in 1086,DB: boc stede, birch-tree place.

BRADFIELD
Bradefelda was held by Roger of Raismes (Nord, France) in 1086,DB: brad feld, broad open country. *(Fig. 6).*

BRADWELL-BY-COGGESHALL
Bradewell, 1238,SR: brad wielle, broad spring/ stream. It lies close to River Blackwater.

BRADWELL-ON-SEA
Odo held *Hacflet* and **Eudo** the Steward held *Acleta* in 1086,DB: haca fleot, angular inlet (?refers to the quay creek). *Bradwell*, 1212,Fees: wide spring. *See Bradwell, St. Peter's Chapel. (Fig. 7).*

BRAIN RIVER
aqua de Wyham, 1254,Ass. *Brain* river, 1848, White's map. A back-formation from Braintree.

BRAINTREE
Ledmar of Hempstead annexed *Branchetreu* in 1086,DB: treow, Branca's tree or wooden-cross.

BRAXTED

Eudo the Steward held *Bracteda* in 1086,DB: bracu or bræc stede, fern-covered or newly-cultivated place.

BRENTWOOD

Bosco arso, 1176,P; *Brendwode,* 1254,Ass: Latin boscum, wood; Middle English brende, burned wood.

BRIGHTLINGSEA

Brictriceseia was held by King **William** in 1086,DB. William II confirmed *Brihtlingeseya* to St. John's Abbey Colchester, *c.*1098,Colch: eg, **Beorhtric's** island or low-lying land.

BROMLEY

Ralph Pinel, a king's burgess, held *Brubeleia* in 1086,DB: brom leah, broom clearing.

BROOMFIELD

Geoffrey de Mandeville held *Brumfelda* in 1086,DB: brom feld, broom open country.

BROXTED

*c.*1050, (K907), King **Edward** confirmed the grant of *Brocheseue* to Ely monastery. After Richard son of Count Gilbert had taken it, *Brochesheued* was restored to Ely *c.*1080,LE. In 1086,DB, Ely held 3 hides and **Eudo** held 9 acres in *Broc(c)hesheuot.* Ely held *Broces hæfd c.*1090,IE: brocc heafod, badger's head (perhaps a reference to the shape of the hill above a tributary of River Chelmer, or Dr. Reaney suggests a place of animal sacrifice). *See Chaureth.*

BRUNDON, BALLINGDON

Ralph de Limésy (Seine Maritime) held *Brumduna* in 1086,DB: brame dun, bramble hill. *(Fig. 6).*

BUCKHURST HILL

Bocherst, 1135,BL: boc hyrst, beech-wood hill.

BULMER

Bulenemera was held by **Richard** son of Count Gilbert in 1086,DB: bula mere, pool where bulls drink.

BULPHAN

St. Mary's, Barking, held *Bulgeuen* in 1086,DB: burh fenn, fen with fortification. (The fortification site is as yet not identified but was perhaps on higher ground near the manor house. There is a moat at The Downes). *(See **Erkenwald**).*

BURES

Richard son of Count Gilbert held *Buro, Bura* in 1086,DB: bur, dwellings. *See also Mount Bures.*

BURNHAM

Ralph Baynard held *Burneham* in 1086,DB: burna ham, settlement by the stream. A tributary of Pannel's Brook flows by the Hall north of the present village which now lies on the River Crouch.

BURSTEAD, GREAT, LITTLE

*c.*989,(CS 1306), **Ælfhelm**, a minister, confirmed the gift to his wife of *Burgestede.* **Odo** and the Bishop of London respectively held (Great and Little) *Burghesteda,* 1086,DB: burh stede, fortified place. (The exact site of the fortification is at present uncertain but the large Norman church suggests a place of some importance).

BUTTSBURY

Henry of Ferrers held *Cinga* in 1086,DB: people of the district; in 1219,FF, Thomas and Ydonia de Mareseis quitclaimed to Geoffrey fitz Theobald one-third of a marsh called Suthanewurth in *Botolfvespirie,* for 1 mark: pirige, Botolph's pear-tree. *See Fristling Hall.*

CAM RIVER

Styrice, 1004,(K711); *Camus,* 1586,Camden: back-formation from Cantebruge, *c.*1120 (Cambridge). *(See also Granta).*

CAN RIVER

Can River, 1801,OS: ?back-formed from Canfield.

CANEWDON

Carenduna was held by **Swein** in 1086,DB: dun, hill of Cana's people.

CANFIELD
Count **Alan** and William de **Warenne** respectively held (Great and Little) *Canefelda* in 1086,DB. William de **Warenne** I confirmed his father William's grant of the church at *Kaneuelde* to Lewes priory *c.*1095 (ECE 109). William II freed the monks of royal dues from *Kanefelda* church *c.*1095 (ECE 118): feld, Cana's open country.

CANVEY ISLAND
In 1254,FF, Everard and Agnes de Schobr' acknowledged the right of Luke de Abbingeton to 1 messuage and 37 acres in ... and *Caneveye*: eg, island of Cana's people. ("Canvey Island" is a tautology.)

CASTLE POINT
Administrative Unit created 1974 from Benfleet and Canvey: Hadleigh CASTLE/Canvey POINT.

CHADWELL ST. MARY
Celdeuuella was held by the Bishop of London in 1086,DB: cald wielle, cold spring. St Mary's church.

CHAFFORD
Ceford, 1066,WDB; *Ceffeorda, Ceffeworda*, 1086,DB: cealfa weorth, calves' enclosure, which may have been on Chafford Heath, the site of the Saxon meeting-place (moot).

CHALKWELL (near Southend)
Chalkwell, 1336,Fine: chalk-lined spring.

CHAPEL
Britesbrig, 1272,Ass: brycg, **Beorhtric's** bridge (over River Colne); in 1528,FF, John and Joan Spark *et al.* quitclaimed to Thomas Wete *et al.* 1 messuage, 30 acres of land in Magna Tey and *Chapell parisshe* for £40: a chapel was present in the 13th century.

CHELMER RIVER
Beadewan ea, 1062, (K813): ?Celtic river-name; *Chelmer*, 1576, back-formed from Chelmsford *(qv)*. *(Fig. 27). See Baddow.*

CHELMSFORD
Celmeresfort was held by the Bishop of London in 1086,DB: Ceolmær's ford (through the River Can or Chelmer). *(Fig. 28). See Moulsham.*

27. Barnes mill, built c. 1800, and mill pool, from the east, on the River Chelmer.

28. *Chelmsford High Street in 1808, from the junction with Springfield Road.*

CHESTERFORD, GREAT AND LITTLE

***Æthelwardes** ceasterforda*, 1004,LE. **William I** held (Great) *Cestreforda*, **Walter** the Deacon held (Little) *Cestrefort* in 1086,DB: ceaster ford, ford by walled town. *See Emanuel Wood.*

CHICKNEY

Ranulf Peverel held *Cicchenai* in 1086,DB: eg, Cicca's marsh-land.

CHIGNALL

Cinguehella was held by **Robert** Gernon and *Cingehala* was held by **Geoffrey** de Mandeville in 1086,DB: Cica's or cicen halh, chicken corner.

CHIGNALL ST. JAMES

Chichehal' Sci Jacobi, 1254,Ass. Church dedicated to St. James.

CHIGNALL SMEALEY

Smetheleye, 1254,For: smethe leah, smooth clearing in woods. *(See also above).*

CHIGWELL

Ralph de Limésy (Seine Maritime, *Fig. 6*) held *Cingheuuella* in 1086,DB: Cicca's or ceacge wielle, gorse-bush stream (?the brook which flows west beside Courtland Drive to River Roding). *Durolito* appears in an itinerary of the 3rd cent: (Celtic) duro ritu, walled settlement at the ford, may be at Chigwell or Romford. *See Barringtons.*

CHILDERDITCH

King **William** held *Ciltendis* in 1086,DB: Celta (name in *c*.687 for a boundary of Barking monastery land, Mar Dyke stream) + dic, dyke, bank, ditch.

CHINGFORD

c.1050, (K913), King **Edward** confirmed the grant of *Cingeford* to St. Paul's which held *Cinghefort* in 1086,DB: chingel ford, shingley ford. (There was a ford through the River Lea to the northwest called Dells Ford, now Valley Road, and a ferry, Cooks Ferry to the southwest, now Lea Valley

Viaduct. It is also possible that the ford was through River Ching at Chingford Lane).

CHIPPING ONGAR
Count **Eustace** held *Angra*, 1086,DB: cieping, market, anger, grassland. *See also Ongar.*

CHISHILL
Cishella was held by **Geoffrey** de Mandeville in 1086,DB: cis hyll, gravel hill.

CHRISHALL
In 1068,(ECE 84), **William I** confirmed the grant of *Cristeshala* by Ingleric the priest to St. Martin le Grand, London. It was appropriated by Count **Eustace** who held *Cristeshala* in 1086,DB: halh, Christ's corner.

CLACTON
St. Paul's required 2 men from *Claccingtune* for ship duty, *c*.998, (ECE 28). Bishop of London held *Clachintuna* before and in 1086,DB: tun, bird-sound settlement?

CLAVERING
Clæfring, *c*.1000, Hatfield data on serfs (K1354). **Swein** held *Clauelinga* in 1086,DB: clæfre ing, clover-growing place.

COBBINS BROOK
Lobbing', 1228,Ass: personal name Cobba's place.

COGGESHALL
c.1050,(ECE 52), **Wulfgyth**, widow of Ælfwine, and **Godwine**, ?of Warley, her son, granted *Coggashæle* to Christ Church Canterbury. *c*.1060,(ECE 66), **Ketel** willed *Kockeshale* to his brother Godric. The monastery held (Little) *Coghessala*, Count **Eustace** held *Cogheshala* in 1086,DB: Cocc's or cocc, hillock or rooster + halh, corner. *(Fig. 29).*

COLCHESTER
Camul(oduno), *c*.20 AD,coins: (Celtic) duno, fort of god Camulos (was at the British site at Gosbecks); *Cair Colun*, *c*.800, *Colneceastre*, 921,ASC. Bishop of London held 14 houses and land in *Colecestra*, 1086,DB: (Celtic) cair, ceaster, fort by (River) Colne (refers to the Roman fort bounded to the west by Balkerne Hill).

COLD NORTON
Nortuna was held by **Ralph** Baynard in 1086,DB: tun, north settlement.

29. Coggeshall, the road to Kelvedon.

COLNE
(Engaine) *c.*950,(CS,1012), ealdorman **Ælfgar** willed *Colne* to his younger daughter [**Ælfflæd**]: Celtic or pre-Celtic river-name. **Robert Malet** held *Parva Colun*, 1086,DB; Vital **Engaine** was granted 40 acres in *Colun* by John fitz Robert to hold of the prior and monks of St. Faith of Horsefort for 2/6d rent in 1218,FF; *Colum Engayne*, 1254,Val.

COLNE RIVER
Colnewater, 1256,Paper: Pre-Celtic or Celtic river-name. *See also Colchester.*

COPFORD
*c.*995,(K699), **Æthelric** willed *Coppanforde* to St. Paul's. The bishop held *Copeforda* in 1086,DB: Coppa's ford (Roman river).

CORRINGHAM
Maurice, Bishop of London held *Currincham* in 1086,DB: ham, settlement of Curra's people.

CRANHAM
Odo held *Craohv*, and Maurice, Bishop of London held *Wocheduna* in 1086,DB: crawe hoh, crow ridge *(and see Ockendon).*

30. All Saints, Cressing, 13th-century church.

CREEKSEA
Criccheseia was held by **Odo** in 1086,DB: ciric hyth, hythe (on River Crouch) by the hill.

CRESSING
Cressyng, 1136,St John: cærse ing, place where water-cress grows. *(Fig. 30).*

CROUCH RIVER
Huolne, c1200,Gerv: (Middle English) wholve, arch, concave; *Crouch*, 1576,Saxton map: ?cruc, cross.

DAGENHAM
*c.*687,(CS 87), **Hodilredus**, a kinsman of King **Sebbi**, granted *Dæccanhaam* to Abbess **Ethelburga** for Barking monastery. William de *Dakeham* held 1 virgate from Thomas & Alice de Seppeleg, rental 4/4d in 1248,FF: ham, Dæcca's settlement.

DANBURY
Danengeberia was held by **Geoffrey** de Mandeville in 1086,DB: burh, fort of people of Dæne. Danbury church is at the centre of the site of the Iron Age fort.

DEBDEN, HALL, GREEN, Loughton
Tippedene was granted to Waltham Holy Cross by **Harold** in 1062 (K813). The canons held Tippedana in 1086,DB: Tippa's denu, valley.

DEBDEN, UTTLESFORD
Ranulf Peverel held *Deppedana* in 1086,DB: deop denu, deep valley. Debden Water flows west into River Cam.

DEDHAM
Delham was held by Roger of Raismes (Nord, France) in 1086,DB: ham, Dydda's settlement. *(Fig. 6, Fig. 31).*

DENGIE
*c.*709,(ECE 7), King **Suebred** granted *Deningei* to Ingwald, bishop of London. Bishop Theodred willed *Denesige* to St. Paul's, *c.*950,(CS 1008). **William I** granted 2 hides in *Danseyam* to **St. Valéry** Abbey in 1068,(ECE 85). **Odo** held 2½

31. *The 15th-century perpendicular style church at Dedham, shown on the village sign.*

hides in *Daneseia* in 1086,DB: eg, Daeni's peoples' marshy land. *(See Danbury). (Fig. 6).*

DODDINGHURST
Doddenhenc was held by Robert son of Corbucion in 1086,DB: hyrst, wooded hillock of Dudda's people.

DOMSEY BROOK, EASTHORPE
cp. *Dolwineshey*, 1250,CottCh: associated with Walter Dolfin of Easthorpe, 1291,For: gehæg, enclosure (farm).

DONYLAND
*c.*980,(CS 1288), **Æthelflæd**, widow of King Edmund, willed *Dunninclande* to Ælfwold her priest. Count **Eustace** held (East) *Dunulanda, Dunilanda* in 1086,DB: Dunning's land.

DOVERCOURT
*c.*1000,(CS 1289), **Ælfflæd**, widow of ealdorman **Brihtnoth**, willed *Douorcortae* to King **Æthelred**. **Aubrey** de Vere held *Druurecurt* in 1086,DB: (Celtic) dubr, water + (Latin) cohort, yard or land cut off by water.

DOWNHAM
Dunham, 1168,RBE: dun ham, hill settlement. There is an Iron Age hillfort on the hill-spur.

DUNMOW, GREAT, LITTLE
*c.*950,(CS 1008), Theodred, bishop of London, willed *Dunemowe* to St. Paul's. St. Paul's required 1 man from *Dunmæan* and *Tollesfuntan* for ship duty *c.*998 (ECE 28). **Thurstan** willed (Little) *Dunmawe* to Meruyn *c.*1045,(ECE 59). In 1086,DB, Richard son of Corbucion held (Great) *Dommauua*, **Ralph** Baynard held (Little) *Dommauua* (where he founded the priory): dun mawe, meadow on a hill. *See Bigods, Marks and Southall. (Fig. 32).*

DUNTON (WAYLETTS)
Odo held *Dantuna* in 1086,DB: dun tun, hill settlement. *Weylete*, 1280,Bart: weg-gelaetu, cross-roads.

EARLS COLNE
Aubrey de Vere held *Colne* in 1086,DB. *Aubrey's Colum*, 1141,Colne; *Erlescolne*, 1358,Pat. The Earls of Oxford inherited it. Colne, Celtic or pre-Celtic river name. *See Colne river and Colchester. (Fig. 33).*

EASTER
*c.*1025,LE, **Godgifu**, widow of a Saxon ealdorman, willed *Æstre* to Ely monastery. **Esgar** took (High) *Estra* before 1066, then **Geoffrey** de Mandeville held it, 1086,DB: eowestre, sheepfold. *See Barricks.*

EASTER, GOOD
Count **Eustace** gave *Estra* to St. Martin le Grand, London before 1086,DB. *See* ***Godgifu*** *(above)* and Mashbury.

EASTHORPE
Estorp was held by Count **Eustace** in 1086,DB: thorp, east village. *(Fig. 34).*

EASTON, GREAT, LITTLE
William of **Warenne** and Matthew of Mortagne (La Manche) held (Great) *Estanes*, and **Walter** the Deacon held (Little) *Eistanes* in 1086,DB: stan, Æga's stone.

32. *Chancel chapel of Little Dunmow Priory, founded 1106, in 1786.*

33. *Effigy of Robert de Vere moved from Earls Colne to Bures in 1935.*

34. St Edmund's Norman church, Easthorpe.

EASTWOOD

Swein held *Nestuda, Estuuda* in 1086,DB: wudu, eastern wood.

ELMDON

Elmduna was held by Count **Eustace** in 1086,DB: elmen dun, hill of elm-trees.

ELMSTEAD

Swein held *Almesteda* in 1086,DB: elm stede, elm place.

ELSENHAM

Robert Gernon held *Elsenham, (Alsenham)* in 1086,DB: Elesa's or Alesa's settlement.

EPPING

Count **Alan** held *Eppinga* in 1086,DB: yppe ingas, upland or look-out dwellers.

ESSEX

*c.*675,(K35), Sebbi rex *east Saxonum* witnessed a grant of land to the abbess of Barking; *Prouincia Orientalium Saxonum, c.*730, **Bede**, who reported that in 604 Mellitus was appointed 1st bishop of London and the East Saxons. *(Fig. 35).*

FAIRSTEAD

Fairsteda was held by **Ranulf** Peverel in 1086,DB: fæger stede, pleasant place.

35. George Bickham's bird's-eye 'View of Essex from the Thames', 1754.

FAMBRIDGE

*c.*1025,LE, **Godgifu**, widow of an ealdorman, willed (South) *Fanbrege* to Ely monastery. Reginald Gunner held from the king (South) *Phenbruge* in 1086,DB but Ely Abbey claimed *Fambrug(g)e, c.*1090, IE. Theodoric Pointel held (North) *Fanbruge* in 1086,DB: fenn brycg, bridge by the marsh (by River Crouch). *See Easter for Godgifu.*

FARNHAM

Hamo the king's Steward held *Phernham* in 1086,DB: fearn ham, fern settlement.

FAULKBOURNE

Falcheburna was held by **Hamo** the king's Steward in 1086,DB: fealca burna, falcon stream (River Brain or a tributary running near the Hall).

36. *Holy Cross 14th-century church and 12th-century tower at Felsted.*

FEERING
*c.*1070,WDB, **William I** exchanged *Feringes* for Windsor with Westminster monastery which held *Pheringa* in 1086,DB: ge-fera, Fera's or companion peoples. *See also Ockendon.*

FELSTED
In 1082,(ECE 92), **William I** and his wife Matilda granted *Feldestede* to the nunnery of Holy Trinity Caen in France, which held *Phensteda* and Roger God-save-the-Ladies held *Felstede* in 1086,DB: feld stede, site in open country. *(Fig.* **36***).*

FINCHINGFIELD
King **William** held *Phincinghefelda* in 1086,DB: feld, open country of Finc's people. *(Fig. 37).*

FINGRINGHOE
*c.*980,(CS 1288), **Æthelflæd**, widow of King Edmund, willed *Fingringaho* to ealdorman **Brihtnoth** and her sister [**Ælfflæd**] who willed it *c.*1000 to (West) Mersea church (CS 1289). King Edward granted (West) Mersea including *Fyngeryngho* to St. Ouen (Rouen) monastery, *c.*1046 (ECE 63): hoh, people on the finger-shaped spur?

37. *The green and Norman tower of St John's church, Finchingfield.*

38. *Fobbing marsh and 14th-century church with a Saxon north window.*

FOBBING

In 1068, **William I** confirmed a grant of land in *Fobbinge* by Ingelric the priest to St. Martin le Grand, London. Count **Eustace** held *Phobinge* in 1086,DB: ing, Fobba's place. *(Fig. 38).*

FORDHAM

Fordeham was held by William of **Warenne** in 1086,DB: hamm or ham, river-meadow or settlement by the ford (River Colne).

FOULNESS

*Fulenesse, c.*1190,Bodl: fugol næss, wild birds' promontory. *(Fig. 39).*

FOXEARTH

Focsearde was held by **Richard** son of Count Gilbert in 1086,DB: eorthe, fox-hole.

FRATING

*c.*1060,(ECE 66), thegn **Ketel** left *Fretinge* in his Will. **Ranulf** Peverel held *Fretinga* in 1086,DB: ingas, Fræte's peoples' settlement.

FRINTON

Count **Eustace** held *Frientuna* in 1086,DB: tun, settlement of Fritha or frithen, protected settlement.

FRYERNING

Robert Gernon held *Inga* in 1086,DB: people of the district. In 12th cent. it belonged to brothers of St. John Hospitallers.

FULFEN SLADE, SAFFRON WALDEN

A tributary of River Cam. *Fulefen,* 1248,Ch: ful fenn slæd, foul marshy valley.

FYFIELD

Count **Eustace** held *Fifhida* in 1086,DB: five hides (land tax measurement). *(Fig. 40).*

39. *The sea wall at Fisherman's End, Foulness Island.*

40. *The water mill and mill pool at Fyfield, River Roding.*

GESTINGTHORPE

c.989,(CS1306), **Ælfhelm**, minister to King **Edgar**, willed *Grystlingathorpe* to his wife. **Leofgifu** willed *Gristlyngthorp* to Bury St. Edmunds Abbey *c*.1044, (K931). **Richard** son of Count Gilbert held *Ghestingetorp* in 1086,DB: thorpe, farm of Grystel's people.

GOLDHANGER

Count **Eustace** held *Goldhangra* in 1086,DB: golde hangra, marigold-covered woodland slope.

GOSFIELD

In 1198,FF, Walkelin and Emma de Bedigfelde quitclaimed 50 acres of wood in *Gosfeld* by mort d'ancestre: gos feld, open country with geese. *(Fig. 9).*

GRANTA RIVER

Gronte fluminis, *c*.745,Felix; *Grantan* stream, *c*.890,Bede: gronna (Celtic), muddy. *See Cam.*

GRAYS THURROCK

William Peverel held *Turroca* in 1086,DB: thurroc, ship-bilge; In 1194,FF, Henry de Grai (of the family from Graye, Normandy) claimed the manor from Josce son of Isaac the Jew whose father had bought it from Earl of Derbyshire who had inherited it from the Domesday holder William Peverel of Nottingham.

GREENSTEAD-BY-COLCHESTER

c.950,(CS 1012), ealdorman **Ælfgar** willed *Grenstede* to Stoke-by-Nayland (Suffolk) [church]. Before 1066,DB, Godric held 4 hides at *Grenesteda*: stede, green place.

GREENSTED-BY-ONGAR

Hamo the king's Steward held *Gernesteda* in 1086,DB: grene stede, green place. *(Fig. 41).*

HACKNEY

Robert de Maringney quitclaimed to Roger bishop of London 1 caracute of land in *Hakeney*, 1231,FF: Haca's or hæcc, hatch or haca, river-bend; ea, with marshland.

HADLEIGH

c.998,(ECE 28), St. Paul's required 1 man for ship duty from *Hæthlege*. The canons held *Lea* in 1086,DB: haeth leah, heathy clearing.

HADSTOCK

In 1008, (K725), King **Æthelred** sold *Caddanno* for £9 in gold from Abbot Ælfsige to Ely monastery: hoh, Cadda's promontory. *c*.1050, King **Edward** confirmed the holding of *Hadestoc* by Ely (K907). Headda's stoc, (?religious) place. Ely held *Cadenhou* in and before 1086,DB.

HALLINGBURY

Halingeb(er)ia was held by Maurice, Bishop of London in 1086,DB: burh, Healla's people's fort, (i.e. Wallbury Iron Age fort). *(Fig. 42).*

HALSTEAD

William de **Warenne** held *Halsteda* in 1086,DB: hald stede, place of shelter.

HAM, EAST, WEST

In 958,(CS 1037), King **Edgar** granted 5 little houses in *Hamme* to ealdorman **Æthelstan** and in 969 confirmed a grant of *Hamme* to Westminster monastery in 969,(CS 1264). In 1086,DB, Westminster held 2 hides in (East) *Hame* and **Robert Gernon** held 15 hides 30 acres in (East and West) *Hame* with **Ranulf Peverel**: hamm, flat land on river bend.

41. *The 11th-century St Andrew's church, Greensted-by-Ongar.*

42. *The enlarged 16th-century Hallingbury Place, Great Hallingbury, 1819.*

HANNINGFIELD
*c.*1036,LE, **Lustwine**, father of **Thurstan**, willed (South) *Hamniggefelde* to Ely monastery. **Odo** held (South) *Haneghefelda*, **Ralph** Baynard held (East) *Haningefelda*, William de **Warenne** held (West) *Haningefelda*, in 1086,DB, but Ely claimed (South) *Hammingefelda, Hanningafelda, Hamugefeld*, *c.*1090,IE: ing, open country of Hana's people.

HARLOW
*c.*1045,(ECE 59), **Thurstan**, son of **Lustwine**, willed *Herlawe* to Bury St. Edmunds monastery which held *Herlaua* in 1086,DB: here hlaw, army hill/mound (probably at the Romano-British temple which is surrounded by an earthwork).

HAROLD WOOD
Horalds Wood, *c.*1237,HPD: Wood of Earl **Harold Godwinson** who owned Havering manor before 1066,DB.

HARWICH
Herdwic, 1229,Pat: herewic, (Viking) army camp.

HATFIELD BROAD OAK (Regis)
*c.*1010,LE, **Leofwine** son of Æthulf granted the annual farm of the king's vill at *Æthfelda* to Ely monastery. King **Æthelred** granted 20 hides at *Nadfelda* to Sigered [his minister] in 1013 (Kelly, 931a). *Hadfelda* was held by King **William** and Hamo the Steward held ½ hide Siriceslea [Ryes] in *Hatfelde* in 1086,DB: hæth feld, heathy open country. (A large oak in Hatfield Forest).

HATFIELD PEVEREL
*c.*1000,(CS 1289), **Ælflæd**, widow of ealdorman **Brihtnoth**, willed *Hæthfelda* to Stoke-by-Nayland (Suffolk) [religous community]: hæth feld, heathy open country. **Ranulf** Peverel was holder of *Hadfelda* in 1086,DB.

HAVENGORE
Havenemersche, 1230,FF: hæfen mersc gara, marshy harbour triangle (at the mouth of the Thames).

HAVERING-ATTE-BOWER
King **William** held *Haueringas* in 1086,DB: Hæfer's peoples; *atte-Bower*, 1272,Pat: bur, at the (king's) residence. The Domesday manor included Romford and Hornchurch.

HAWKWELL
Eudo the Steward held *Hechuuella* in 1086,DB: haca wielle, angled stream (tributary of R. Roach).

HAZELEIGH
Ranulf Peverel was the holder of *Halesleia* in 1086,DB: leah, Hægel's clearing.

HEDINGHAM, CASTLE, SIBLE
Aubrey de Vere and Roger of Raismes (Nord) held *Haingheham, Hedingham* in 1086,DB: hyth ham, settlement of people at a landing-place (on River Colne). *(Fig. 6)*. The castle was built *c.*1135 by Aubrey's son. *Sibil*, widow of Geoffrey de Laventon, in 1237,FF, sold 6 acres of land to Robert and Katherine Curpay.

HELION BUMPSTEAD
*c.*1045,(ECE 60), Wulfwin, son of Abbot Alfwin, granted *Bumestede* to Ramsey Abbey, Cambs. **Aubrey** de Vere and Tihel de Helléan (from Morbihan) held *Bumesteda, Bunsteda* in 1086,DB: ? bune stede, reedy place. *(Fig. 6)*.

HEMPSTEAD
Richard son of Count Gilbert held *Hamsteda* in 1086,DB: henep stede, hemp place or hamstede, homestead.

HENHAM
*c.*1045,(ECE 59), **Thurstan**, son of **Lustwine**, willed *Henham* to his wife Æthelgyth but **Ralph** Baynard held *Henham* in 1086,DB: heah ham, to the high settlement.

HENNY, GREAT, LITTLE
Ranulf Peverel held *Henies* in 1086,DB: henn or heah eg, bird or high island (at 71m above tributaries of River Stour).

HEYBRIDGE, MALDON

*c.*950,(CS 1012), ealdorman **Ælfgar** willed *Tidwoldingtone alias Tidweldington* to Ælfwold. *c.*1000, (CS 1289) **Ælfflæd** willed *Tidwoldingtune* to St. Paul's which held *Tidwolditunа* in 1086,DB: tun, Tidweald's settlement. *Heaghbregge*, *c.*1200,Gerv: heah brycg, high bridge (above River Blackwater).

HIGH RODING

*c.*1090,(ECE 108), William de **Warenne** granted the church at *Roynges* to Lewes priory. *Roinges Doun' Bard'*, 1194,P: Hrotha's peoples' settlement, Doun Bardulf granted 3 acres to Dunmow Priory. (Sited on the Roman Road at 89m O.D.). *See Roding.*

HOCKLEY

King **Eadred** granted 19 hides at *Hoccanleage* in 946 to a religious woman Eawynne (Kelly 517b). Barking monastery held *Hocheleia* in 1086,DB: hocc(a) leah, Hocca's or mallow clearing.

HOLLAND

*c.*962,LE, **Ædgiva**, widow of King **Edward the Elder,** willed *Holand* (which she bought from Sprowe for £20) to a lady **Ælftred** (see **Ælfthryth**). Count **Eustace** held (Little) *Hoilanda* in 1086,DB: hoh land, land on a promontory (into North Sea).

HORKESLEY, GREAT, LITTLE

Eiland (in the lordship of Nayland, Suffolk) was held by **Swein**, 1086,DB; æt thæm eglande, at the river land. *Horchesleia*, *c.*1130,Bodl: horsc leah, foul clearing/ wood. *(Fig. 43).*

HORNCHURCH

Monasterium Cornutum, 1222,ERxxix. Richard and Margery Newman acknowledged the right of Bro. Turumbertus master of the Hospital and monastery of Hornchurch (*de cornuto*) of 40 acres in Havering, 1242,FF: church with horn-shaped gable?

43. 13th-century wood effigies of the de Horkesleys, Little Horkesley church.

HORNDON, EAST, WEST
King **Æthelred** granted 5 hides in 1013 at (West) *Thorndun* to his minister Sigered (Kelly, 931b). **Odo** and **Swein** held *Torinduna, Tornindun*a in 1086,DB: thornigan dune, hill of thorn bushes.

HORNDON-ON-THE-HILL
Maurice, Bishop of London held *Hornindun*a in 1086,DB: horning dun, horn-shaped hill?

HUTTON
c.1070,(ECE 87) **William I** gifted *How* to found St. Martin's Abbey, Battle (Sussex) which held *Atahov* in 1086,DB: æt hoh, at the promontory hill. The remains of the abbey on the site of the battle in 1066 can still be visited. *(Fig. 44)*.

44. The monks' dormitory, Battle Abbey, Kent, on the site of the Battle of 1066.

ILFORD
Jocelyn Lorimer held *Ilefort* in 1086,DB: ford through river Hile (Celtic sil, 'trickling stream'), now Roding river, *qv*.

INGATESTONE
Inga was held by St. Mary's, Barking in 1086,DB: people of the district. Ginge attestone, 1289,Ch: probably referring to the two sarsens by the crossroads and one at the church. *(front cover)*.

INGRAVE
Odo held *Inga* in 1086,DB: people of the district; *Ginge Rad'*, 1238,Subs: Ralph, son of **Thorold**, was Domesday subtenant in 1086.

INGREBOURNE RIVER
Ingceburne forms the western boundary of Upminster, 1062 (K813): burna, Ine's? stream

INWORTH
Inewrth, 1206,Cur: worth, Ina's homestead.

KELVEDON
Canonio, 3rd century, Antonine Itinerary: (Celtic) cano, reedy river. In 998,(ECE 32), thegn **Leofwine**, son of **Wulfstan** Bishop of London, willed *Cynlaue dyne* to Westminster monastery which held *Chelleuedana* in 1086,DB: denu, Cynelaf's valley, (River Blackwater).

KELVEDON HATCH
c.1043,(ECE 51), King **Edward** confirmed the grant of *Killeuendun* by **Æthelric** the chamberlain and his wife Gode to Westminster monastery which held *Keluenduna* in 1086,DB: cylu dun hæc, (to the) multicoloured hill gate (to former forest).

KIRBY-LE-SOKEN
Kyrkebi, 1181,StP: Old Scandinavian kirkjuby, church village; socn, a district under jurisdiction. *See Walton*.

LAINDON
c.998,(ECE 28), St Paul's required 1 man from *Ligeandune* for ship duty. The bishop of London held *Legenduna* in 1086,DB: dun, hill by the Lyge (Celtic river name). (River Crouch tributaries rise in the vicinity).

LAMARSH
Ranulf Peverel held *Lamers* in 1086,DB: lam ersc, loam ploughed field.

LAMBOURNE
Count **Eustace** was holder of *Lamburna* in 1086,DB: lam burna, loam-stream, or lamb burna, stream where lambs are. A tributary of River Roding runs north from Bishop's Hall.

45. St Michael's 14th-century curch, Latchingdon, in 1780.

LANGDON
Swein held *Langenduna* in 1086,DB: langan dun, long hill (which still well fits the topography).

LANGENHOE
Count **Eustace** held *Langhou* in 1086,DB: langan ho, long promontory (between Geedon Creek and Pyefleet Channel).

LANGFORD
Langefort was held by **Ralph** Baynard in 1086,DB: lang ford, long ford (through River Blackwater).

LANGHAM
Richard son of Count Gilbert held *Laingaham* in 1086,DB: ham, settlement of Lahha's people.

LANGLEY
Langelega, 1166,RBE: leah, long wood/clearing.

LATCHINGDON
In 1065,(K824), King **Edward** confirmed the grant of *Læcedune* by Guthmund to Westminster Abbey. **Hugh** de Montfort held *Lachentuna* and **Wulfeva** wife of **Finn** the Dane held *Lacenduna* in 1086,DB: læccen, snare or læcen dun, well-irrigated hill. Several streams run north into Mundon Creek. *(Fig. 6; Fig. 45).*

LATTON
Bury St. Edmunds Abbey held *Lattuna* in 1086,DB: leac-tun, leek (herb)-garden settlement?

LAVER, HIGH, LITTLE
*c.*1010, (K972), Æthelflæd, sister of ealdorman **Leofsige**, granted *Lagefare* to St. Paul's. Count **Eustace** held *Lagafara* in 1086,DB: lagu fær, water ford (through a tributary of Cripsey Brook).

LAVER, MAGDALEN
Ralph of Tosny (Eure) held *Laghefara* in 1086,DB: St. Mary Magdalen church. *See Laver above.*

LAWFORD
*c.*1040,(K931), **Leofgifu** willed *Lalleford* to Æthelric her household chaplain. **William I** held *Laleforda* in 1086,DB: Lealla's ford. There are several possibilities for the ford site, at White's Bridge through River Stour, through Shir Burn at Mill Hill or through Wignall Brook at Church Hill which is the nearest to the Hall.

46. The gatehouse built c.1520, at Layer Marney.

LAYER BRETON, HAYE, MARNEY
Count **Eustace**, Bishop of London and **Ranulf** Peverel held *Legra* in 1086,DB: Celtic river-name (for Layer Brook), Ralph *Brito* (the Breton) son of Lewis granted St. John's Abbey Colchester 2 parts of tithes in Layer Breton *c.*1300; Maurice de la *Haia* and his son Ralph (died 1253) were benefactors to St. John's Abbey Colchester; in 1166 William de *Marney* held a knight's fee from Henry de Essex. *(Fig. 6, Fig. 46).*

LEA RIVER
The Vikings built a fort by the river *Lygan* in 894, Anglo-Saxon Chronicle: lug (Celtic) bright, or dedicated to Celtic god Lugus.

LEADEN RODING
*c.*1090,(Dudg), William de **Warenne** gave the church at *Roinges voc. Ledenechirche* for his foundation of Castle Acre priory: lead-roofed church. *See Roding.*

LEE CHAPEL
The Patron of *Leye* chapel was the heir of Robert Lee, 1254, TEASxviii,22; EAHxxi,56: leah, wood/clearing.

LEIGH-ON-SEA
Ranulf Peverel held *Legra* in 1086,DB: leah, clearing.

LEIGHS, GREAT, LITTLE
Eudo the Steward and **Geoffrey** de Mandeville held *Lega, Legra*, 1086,DB: leah, clearing. *(Fig. 47).*

LEXDEN
*c.*1000,(CS1289), **Ælfflæd**, widow of ealdorman **Brihtnoth**, willed *Læxadyne* to King **Æthelred**. **William I** held *Lessendena* in 1086,DB: denu, Leaxa's valley.

LEYTON
In 1065,(K824), King **Edward** confirmed **Atsere Swerte**'s ('the swarthy') grant of *Lygetun* to Westminster Abbey. The abbey held *Leituna* in 1086,DB: tun, settlement by river Lea *(qv).*

47. The 16th-century tower gateway, Leighs Priory.

LIMEBROOK, MALDON
Lymborne broke, 1488,Beel: lind burna, lime-tree stream.

LINDSELL
Lindesela was held in 1086,DB, by **St. Valéry** Abbey, France: lind sele, dwelling among lime-trees.

LISTON
*c.*1000, (CS1289), **Ælfflæd**, widow of ealdorman **Brihtnoth**, willed *Lissingtune* to **Æthelmær**. **William I** confirmed the grant by **Hugh** de Gornaco of ½ the tithe of *Liston* (*Listuna*, 1086,DB) to Bec Abbey, 1087 (ECE 100): tun, Lissa's settlement.

LITTLEBURY
*c.*989, (CS1306), **Ælfhelm**, minister to King **Edgar**, willed *Lytlanbyrig* to **Leofsige**, his kinsman. King

48. The Norman south door of the church at Mashbury.

Æthelred granted 20 hides at *Lyttlanbyrig* to Ely monastery in 1004 (K711). Ely held *Litelbyria* in 1086,DB, *Littleberi c.*1090,IE: burh, little fort; This is Ring Hill Iron Age fort.

LOUGHTON

In 1062, (K813), Earl **Harold** granted *Lukintone* to found Waltham Holy Cross. The canons held *Lochintuna* in 1086,DB: Luca's tun, settlement.

MALDON

At *Mældune* in 909, ASC, King **Edward** encamped in preparation to engage the Danes. In 1068, **William I** confirmed the grant of the church at *Mealdone* to St. Martin le Grand, London (ECE 84). He held *Melduna* in 1086,DB: mæl dun, hill with a cross (?before a church was built).

MANNINGTREE

Manitre, 1248,Ass: tre, Mann's or manig, many trees.

MANUDEN

Robert Gernon held *Menghedana* in 1086,DB: denu, Manna's peoples' valley.

MAPLESTEAD, GREAT, LITTLE

In 1065,(K824), King **Edward** confirmed Wulfwin's grant of *Mapulderstede* to Westminster Abbey. **Robert** Gernon held (Great), John son of Waleran held (Little) *Mapledesteda*, 1086,DB: mapuldor stede, maple-tree place.

MAPLIN SANDS

Mablynge Swene, 1421,ERi: first element is uncertain + swin, channel.

MARGARET RODING

Roinges Sce Margar', *c.*1250,Ass: Hrotha's peoples' settlement, church of St. Margaret. *See Roding.*

MARGARETTING

King **William I** held *Ginga* in 1086,DB: people of the district; *Genges Sce Margar'*, 1285, Ass: St. Margaret's church.

MARKSHALL

In 998,(ECE 32), thegn **Leofwine** willed *Mearcyng seollan* to Westminster monastery. **Hugh** de Montfort held *Mercheshala* in 1086,DB: ing sele, Mearc's dwellings or mearc sele, place of boundary dwellings (between Lexden and Hinckford Hundreds).

MASHBURY

In 1068, (ECE 84), **William I** confirmed the grant of *Maisbirig* by Ingelric the priest to St. Martin le Grand, London. Before 1086, Count **Eustace** appropriated an outlier (i.e. part) of Good Easter from St. Martin, and **Geoffrey** de Mandeville held *Massebirig* in 1086,DB: burh, Mæcca's fort (probably at Oldberry field, *Essex Journal* (2005) 40, p14). *(Fig. 48).*

MATCHING

St. Valéry Abbey held *Metcinga* in 1086,DB: Mæcca's peoples' settlement or ?stream-name (for a tributary of Pincey Brook).

MAYLAND

Mailanda, 1188-92,P: mægtha, mayweed, or æt thæm eglande, at the low-lying land.

MERSEA, EAST, WEST

The Viking host gathered on *Meres ig* in 894, ASC. **Ælfgar** willed Totham to **Æthelflæd** and (West) *Mereseye* community *c*.950 (CS 1012). King **Edward** granted (West) *Mersege* to St. Ouen (Rouen) Abbey in 1046 (ECE 63). The abbey held (West) *Meresaia*, **Swein** held (East) *Meresai* in 1086,DB: meres eg, pool by an island.

MESSING

Metcinges was held by **Ralph** Baynard in 1086,DB: ing, Mæcca's peoples' settlement.

MIDDLETON

Richard son of Count Gilbert held *Mildeltuna* in 1086,DB: middle settlement.

MISTLEY

Roger of Raismes (Nord) held *Mitteslea* in 1086,DB: mistel leah, mistletoe wood. *(Fig. 6). See Dickley and Cliff Halls.*

MORETON

William of Écouis (Écouis, Eure) held *Mortuna* in 1086,DB: mor tun, settlement in moorland.

MOUNT BURES

Roger of Raismes (Nord) and **Roger** of Poitou held *Bura* in 1086,DB: bur, cottage, by a mound. *(Fig. 6). See also Bures.*

MOUNTNESSING

Ranulf brother of Ilger held *Ginga* in 1086,DB; Peter Foliot acknowledged that Hugh abbot of Stretfford has of the gift of his father Gilbert 2 marks rent in *Gynges Munteny*, 1236,FF: people of the district, Robert de *Mounteney's* name appears on the foundation charter of Thoby Priory), 1141,M. *(Fig. 49).*

MUCKING

St. Marys, Barking held *Mucinga* in 1086,DB: settlement of Mucca's people or (Germanic) muk, soft, muddy place. *(See **Erkenwald**).*

49. Mountnessing post mill built in 1807 and mill mound.

MUNDON

Eudo the Steward held *Mundana* in 1086,DB: Munda's or mund dun, protected hill (?from floods).

NAVESTOCK

In 867, (CS 1210), King **Edgar** sold *Naesingstoc* for 60 gold mancuses to St. Paul's which held *Nasestoca, Nessetocha* in 1086,DB: næss ingas, dwellers at the promontory place (above River Roding).

NAZEING

c.700, (Kelly,65a,b), King **Suebred** granted 40 manentes of land at *Nasyngum* to 'Fymme' (Bascombe, 1989). In 1062, (K813), Earl **Harold** granted *Nesingan, Nassingan* to found Waltham Holy Cross. The canons held *Nasinga* in 1086,DB: næss ingas, dwellers at the promontory (above River Lea).

NETTESWELL

Nethleswelle, 1062,Th: wielle, Nethel's stream/spring (possibly now Todd Brook).

NEVENDON

Alfward Dore (OE dora, humble-bee) held *Nezendena*, then Roger Marshal in 1086,DB: efnan dena, at the flat valley.

50. The Leper Stone at Newport.

NEWHAM
Newly-formed borough in 1965 from East and West Ham etc *(qv)*.

NEWPORT
King **William** held *Neuport* in 1086,DB: port, new town. *(Fig. 50)*.

NOAK HILL
Noak Hill, 1490,Ct: Middle English atten ac, at the oak.

NORTH WEALD BASSETT
Peter de Valognes *(Fig. 6)* held *Walda, Walla* in 1086,DB: weald, forest land. Grant by Baldwin son of Hugh of Essex of *North Waude* for 100 marks to Sir Philip Basset (died 1271), sheriff of Somerset and Dorset, in 1260.

NORTON MANDEVILLE
In 1068, (ECE 84), **William I** confirmed the grant of *Nortona* by Ingelric the priest to St. Martin le Grand, London: north settlement (north of High and Chipping Ongar); *Nortuna Maundeuil*, 1262,For, Ernulf de *Mandeville* and his wife Galiena sold 80 acres with appurtances in *Norton* in 1251.

NOTLEY, WHITE, BLACK
In 998, (ECE 32), thegn **Leofwine** willed land in Purleigh on the east side of the street to *Hnutlea* [church]. Count **Eustace** held (White), **Geoffrey** de Mandeville held (Black) *Nutlea* in 1086,DB: nut leah, clearing/wood.

OAKLEY, GREAT, LITTLE
Robert Gernon held (Great) *Accleia* and **Ralph** Baynard held (Little) *Ade [Acle]* in 1086,DB: ac leah, oak-tree wood/clearing.

OCKENDON
*c.*1075,WDB, **William I** granted (North) *Wokindon* to Westminster monastery in exchange for Windsor. Westminster held (North) *Wochenduna* in 1086,DB. Turold steward of **Geoffrey** de Mandeville gave 2/3rds of the tithe of (South) *Wochendona* to found Hurley-on-Thames priory in 1080 (ECE 93). Turold held *Wochaduna* from Geoffrey in 1086,DB: dun, Wocca's hill. The area is too wide for certainty but the highest point at 41m is east of White Post Farm. *See also Feering.*

ONGAR, HIGH, CHIPPING
*c.*1045, **Thurstan**, son of **Lustwine** willed the wood at *Aungre* to his servants. **William I** confirmed a grant by Ingelric the priest of certain lands of (Chipping) *Angra* to St. Martin le Grand London, 1068 (ECE 84). John son of Waleran held (High) *Angra*, 1086,DB: anger, grassland. *See Chipping, Ashlyns, Nash, Paslow and Norton Halls.*

ORSETT
In 957, (ECE 14), King **Eadwig** granted *Orseathan* to Brihthelm, bishop of London. *c.*998, (ECE 28), St. Paul's required 1 man for ship duty from *Orseathun*. St. Paul's held *Orseda* in 1086,DB: ora seath, iron-ore pit.

OVINGTON
Ouituna was held by **Roger Bigot** in 1086,DB: ingtun, Ufa's settlement. *See **Bartholomew Bigod**.*

PAGLESHAM
In 1065, (K824), King **Edward** confirmed the grant of *Paclesham* by Ingulf to Westminster Abbey which held *Pachesham* in 1086,DB: ham, Pæccel's settlement.

51. *St John the Baptist 14th-century church, Pebmarsh.*

PANFIELD

*c.*1075, (ECE 32), **William I** confirmed **Waleram** son of Ranulf's grant of *Paunfeld* to St. Stephen's Abbey, Caen which held *Penfelda* in 1086,DB: feld, open country by river Pant.

52. *The 15th-century hall, Pentlow. (Photo courtesy of the owners)*

PANT RIVER

*in ripa Pentae amnis, c.*730,**Bede**: Cedd's church on the bank of (Celtic) the Pant, valley with river. The name applied to the Blackwater river *(qv)*.

PARNDON

Ranulf brother of Ilger held *Perenduna*, 1086,DB: peren dun, pear-tree hill (above Canons Brook).

PATTISWICK

Pateswik', 1235,Ass: wic, Pætti's dwelling/farm.

PEBMARSH

Richard son of Count Gilbert held *Pebenhers* in 1086,DB: ersc, Pybba's ploughed-field. *See Plunker's. (Fig. 51).*

PELDON

*c.*950, (CS 1012), ealdorman **Ælfgar** willed *Piltendone* to **Æthelflæd** then to **Æfflæd** who willed *Peltandune* to Stoke-by-Nayland community (Suffolk) *c.*1002 (CS 1289). **Swein**

and **William the Deacon** held *Peltenduna* in 1086,DB: pyltan dun, Pylta's or thrusting hill which extends southeast towards the salt marshes.

PENTLOW

*c.*1036,LE, **Lustwine** willed *Pentelawe* to his son **Thurstan** thence to his wife Æthelgyth. **Ralph** Baynard held *Pentelauua* in 1086,DB: hlaw, Penta's or River Pant barrow. *(Fig. 52).*

PINCEY BROOK

Wardbroc, a.1272,Colch; Wearda's brook. *Pynkesey*, 1371,Londin; pinca eg, chaffinch low-lying land?

PITSEA

Wulfeva wife of Finn, held *Piceseia* in 1086,DB: ea, Pic's island or low-lying land in marsh.

PLESHEY

*c.*1050,LE, **Wulfric**, abbot of Ely, agreed that **Esgar** the Staller should hold *Plassiz* for his life then revert to Ely, but **Geoffrey** de Mandeville held *Estra* of which Pleshey was part in 1086,DB, still claimed by Ely: plessis (Norman French), fence of living branches. *(Fig. 53).*

PRITTLEWELL, SOUTHEND

Swein held *Pritteuuella* in 1086,DB: pritiol wielle, babbling stream. (Prittewell Brook).

PURLEIGH

In 998, (ECE 32), **Leofwine**, son of **Wulfstan**, willed land to the east of the street at *Purlea* to [White] Notley [church]. Count **Eustace** and **Hugh de Montfort** held *Purlai* in 1086,DB; pur leah, bittern or snipe wood.

QUENDON

Kuenadana was held by **Eudo** the Steward in 1086,DB: cwenena denu, women's valley (R. Cam).

RADWINTER

Eudo the Steward held *Redeuuintra* in 1086,DB: Rædwynn's (a woman's name) + treo, tree or ræd winter, red vineyard.

53. The bridge leading to the Norman castle motte, Pleshey, in 1784.

RAINHAM

Odo held *Reneham*, **Robert** Gernon held *Renaham* in 1086,DB: Regna's or rogian ham, prevailing people's settlement.

RAMSDEN BELLHOUSE

Ramesdana was held by Bishop of London, 1086,DB: hramsa denu, garlic valley. Richard de *Belhus*, Cambridgeshire, 1200,M.

RAMSDEN CRAYS

Odo held *Ramesdana* in 1086,DB: hramsa denu, garlic valley. Simon de *Creye*, in 1257, expelled Andrew de Blund from the manor.

RAVENSBOURNE RIVER, Hornchurch

tributary of Rom/Beam River. *Bowles Brook*, 1777; *Ravens Bridge*, 1777,C&A.

RAWRETH

Raggerea, 1177,P: hraga rith, heron stream. (Tributary of River Crouch).

RAY SAND, DENGIE

Reysand, 1367,Pat: æt thære ege, at the island (now inundated) + sand.

RAYLEIGH

Swein held *Ragheleia* in 1086,DB: ræge leah, roe-deer or she-goat woodland clearing.

RAYNE
*c.*989, (CS 1306), **Ælfhelm** willed *Hraegan* to his wife; *c.*995, (K699), **Æthelric** willed *Raegene* to the bishop of St. Paul's who held *Raines* in 1086,DB: ?river name for Brain river related to O.E, reen, ditch; or hrægene, shelter/eminence place.

REDBRIDGE
*Hocklee's Bridge,*1650,Tasker; *Red Bridge,* 1735,map: built of red brick for vehicles where Redbridge Lane crossed river Roding, demolished 1759. Nearby White Bridge of wood, a foot-bridge. Since 1965, borough amalgamating Wanstead, Woodford, Ilford, parts of Chigwell and Dagenham.

RETTENDON
*c.*991,LE, ealdorman **Brihtnoth** willed *Ratendune* and *c.*1002, (CS 1289), his widow **Ælffæd** willed *Rettendune* to Ely monastery which held *Rattenduna* in 1086,DB: rætt dun, rat-infested hill. *(Fig. 54).*

RICKLING
Richelinga was held by King **William** in 1086,DB: (Queen?) **Ricola's** peoples' settlement.

RIDGEWELL
Count **Eustace** held *Rideuuella* in 1086,DB: Rædel's or hreod, reedy, wielle spring or stream. (A headwater of River Colne).

RIVENHALL
In 1068, (ECE 84), **William I** confirmed the grant of privileges in *Reuenhala* by Ingelric the priest to St. Martin le Grand, London. Count **Eustace** held *Ruenhale* in 1086,DB: hreofan halh, rough sheltered place. *(Fig. 55).*

ROACH RIVER
Walflestreme, 1374,Ct: wall fleot, embanked estuary.

ROCHFORD
Swein held *Rochefort* in 1086,DB: ræcces ford, hunting-dog ford (through a tributary of River Crouch).

54. Hyde Hall, Rettendon.

55. Repton's proposals for improvements at 16th-century Rivenhall Place, 1794.

RODING *(see separate parishes)*
*c.*1010,LE, **Leofwine**, son of Æthulf, granted *Rothinges* to Ely monastery: Hrotha's people's settlement.

RODING RIVER
In 958, (S676), *ealden Hilæ* formed the eastern boundary of East/West Ham: eald, old + (Celtic) hil, trickling water; *Rodon*, 1576 map, back-formed from High Roding etc *(qv)*.

ROMAN RIVER
*bricsfleot, c.*1046,(ECE 63): ?brycg fleot, bridged inlet; *Romyne*, 1338,Ct: from family of John Romayn of Colchester, 1377,Ct.

ROMFORD
Romfort, 1177, TEASvi, rum ford, wide ford (through River Rom). *(See also Chigwell)*.

ROM RIVER
Bourne, 1297,Wm; *Bourne Brook*, 1777,C&A: burna, stream; later river name back-formed from Romford *qv*. The lower part is Beam river, *qv*.

ROXWELL
Rokeswelle, 1291,Tax: wielle, Hroc's or rook stream. (Roxwell Brook, a tributary of River Can).

ROYDON
Ranulf brother of Ilger held *Ruindune* in 1086,DB: rygenan dun, rye hill (above River Stort).

RUNNING WATER BROOK
marcdice, 1062, (K813): mearc dic, boundary dyke (between Upminster and Aveley, south of Cockhide).

RUNWELL
*c.*970, (CS 737), King **Æthelstan** confirmed the grant of *Runweolla* to St. Paul's. The canons held *Runewella* in 1086,DB: hruna or run or ryne wielle, log or council ('wishing-well') or ditch stream. (Tributary of River Crouch).

SAFFRON WALDEN
*Wealadene, c.*1000, Hatfield data on serfs (K1354). **Geoffrey** de Mandeville held *Waledana* in 1086,DB: weala denu, Britons' valley. *Saffornewalden*, 1582,AD: saffron (from Arabic *safran*) cultivation first recorded here in 1545. *(Fig. 56)*.

56. *The 12th-century castle keep, Saffron Walden.*

ST. LAWRENCE alias NEWLAND
Christ Church Canterbury held *Newilant* in 1086,DB: neue land, newly cleared land. *Villa Sancti Laurentii*, 1235,Cl: from the church dedication to **Lawrence**.

ST. OSYTH CHICHE
*c.*950,(CS 1008), Theodred, bishop of London, willed *Cic* to St. Paul's; *Ticc*, *c.*1000. **Wulfgyth**, sister of thegn Eadwine, gave ½ a pound of money to *Seynte* **Osithe**, 1046,(ECE 64). The bishop held *Cice* in 1086,DB: ?Scandinavian kik, church, or keikr, sinuous (?referring to the creek). *(Fig. 57).*

SALCOTT
Saltcot, 1199-1216,BL: salt cottages: Site of salt-making from sea water.

SALING
John son of Waleran held *Salinges* in 1086,DB: sealh, settlement of the dwellers by the willows.

SAMPFORD, GREAT, LITTLE
King **William** held *Sanfort* in 1086,DB: sandy ford (through River Pant).

SANDON
In 1199,FF, Cecily fitz Gilbert quitclaimed to Therieus de Tillingeham 10/0d in rent in *Sandun* for 100/0d: sand dun, sandy hill (between River Chelmer and Sandon Brook).

57. *St. Osyth Priory and later Abbey buildings.*

SHALFORD
King **William** held *Celdeforda* in 1086,DB: sceald, shallow ford (through River Pant).

SHEERING
Peter of Valognes *(Fig. 6)* held *Sceringa* in 1086,DB: Sceri's or OE scearu, boundary or Old German scara, army-peoples' settlement. Sheering's western boundary is the county boundary; it is also within a few miles of Harlow, qv, another 'army' place-name.

SHELLEY
Senleia was held by **Geoffrey** de Mandeville in 1086,DB: scylf leah, clearing on a slope.

SHELLOW BOWELLS
Lambert de *Buella* (Seine Inférieure, *Fig. 6*) held *Scelga* in 1086,DB, from **Geoffrey** de Mandeville: sceolh, winding (river Roding).

SHENFIELD
Count **Eustace** held *Scenefelda* in 1086,DB: sciene feld, fair open country.

SHOEBURY
The Viking army encamped at *Sceobyrig* and made a fort in 894, Anglo-Saxon Chronicle. **Odo** and **Swein** held *Sobia* and *Essobia* in 1086,DB: sceo burh, fort at shoe-shaped spur of land or shelter.

SHOPLAND
In 946,(ECE 12), King **Eadred** granted *Scopingland* to Eawynn, a nun. St. Paul's required 1 man for ship duty from *Seopinglande* with West Orsett *c.*998 (ECE 28). Count **Eustace** held *Scopelanda* in 1086,DB: sceoppa land, newly cultivated land with a building.

SNOREHAM
Snorham, 1238,SR: snor, ?brushwood, ham or hamm, settlement or enclosure.

SOUTH WEALD
In 1062, (K813), Earl **Harold** gave *Welde* to found Waltham Holy Cross whose canons held *Welda* in 1086,DB: weald, woodland. *(See **Tofig**).*

SOUTHCHURCH
In 823 Lifstanus (Leofstan) granted Southchurch to Christ Church Canterbury; *c.*1050, King **Edward** confirmed the grant of *Suthcyrcean* (ECE 8,54). The monks held *Sudcerca* in 1086,DB: cirice, church (in the south of Rochford Hundred).

SOUTHEND-ON-SEA
Stratende, 1309,Ext: stræt, end of the paved road; *venella vocat Sowthende de Pritwell*, 1481,Will: lane called Southend of Prittlewell parish. The first cottages for oystermen began *c.*1767. *(Fig. 58).*

SOUTHMINSTER
*c.*998,(ECE 28), St. Paul's required 5 men for ship duty from *Suthmynster.* The bishop of London held *Sudmunstra* in 1086,DB: mynster, south minster (south of Cedd's minster at Bradwell Ythancæstir, *qv.*), it having been restored by **William I** to St. Paul's after being taken away by Cnut (king 1016-1035).

SPRINGFIELD
Springinghefelda was held by **Robert** Gernon in 1086,DB: spring feld, open country with copses.

STAMBOURNE
Geoffrey de Mandeville had annexed *Stanburna* before 1086,DB: stan burna, stoney stream. (A headwater of River Colne.)

STAMBRIDGE, GREAT, LITTLE
*c.*1072, (ECE 89), **William I** restored (Little) *Stanbrigge* to Christ Church Canterbury which held *Stanbruge* while **Odo** held (Great) *Stanbruge* in 1086,DB: stan brycg, stone bridge (over a stream which flows into Bartonhall Creek, River Roach). *(See **Hamo**).* (Like Sampford, a late change of 'n' to 'm').

58. Southend beach and terrace c.1820.

STANFORD RIVERS

In 1068, **William I** confirmed a grant of privileges in *Stanford* by Ingelric the priest to St. Martin le Grand (ECE 84). Count **Eustace** held *Stanfort,* 1086,DB: stan ford, stoney ford (through a headwater of River Roding), Richard de Ripariis, 1213.

STANFORD-LE-HOPE

In 1361,FF, Robert Borenare of Bradewell quitclaimed to John Cogger of *Stanford 'in the Hope'* 1 messuage, 140 acres for 100 marks: stan ford hop, stony ford by the inlet/valley. (A stream now under London Road near the station from here flows into Mucking Creek).

STANSTED MOUNTFITCHET

Robert Gernon held *Stanesteda* in 1086,DB: stan stede, stoney place. Richard *Mountfitchet* (died 1203), of the house of Monfiquet, inherited the estate.

STANWAY

*c.*1000,(CS 1289), **Ælfflæd** willed *Stanwægun* to King **Æthelred**. **William** I held *Stanewega* in 1086,DB: stan weg, stone paved way (old Roman road which probably followed an Iron Age track).

STAPLEFORD ABBOTTS

Bury St. Edmunds Abbey held *Stapleforda* in 1086,DB: stapol ford, ford at the pillar. It belonged to the *abbot* of Bury St. Edmunds. The ford presumably predated Passingford Bridge. *(Fig. 17, Fig. 59).*

59. Albyns, built c.1620 at Stapleford Abbots, in 1806.

STAPLEFORD TAWNEY

Swein held *Staplefort* in 1086,DB: stapol ford, ford at the pillar *(see Stapleford Abbotts)*. Margaret daughter of William FitzRichard married Richard de *Tany* who in 1253 was granted licence to hunt in Waltham Forest, and in 1264 to enclose Stapleford Tawney wood.

STEBBING

Henry of Ferrers held *Stibinga* in 1086,DB: stybb, settlement of Stybba's people or near the stubbs.

STEEPLE

Stepla was held by King **William** in 1086,DB: stiepel, tower.

STEEPLE BUMPSTEAD

Count **Eustace** held *Bummesteda* in 1086,DB: ?bune, reed, stede, place. *Stepelbumstede*, 1267; *Bumsted at towre*, Norden, 1594. *(Fig. 10)*.

STIFFORD

Barking monastery held *Stiforda* before 1066, DB; Gilebeard held 72 acres in *Stithford*, *c*.1090,(ECE 106): stithe or stith ford, lamb's-cress or difficult ford (through The Mardyke stream). *(Fig. 60)*. *(See* ***Erkenwald****)*.

STISTED

c.1050, **Wulfgyth**, widow of Ælfwine, and **Godwine**, ?of Warley, her son, granted *Stigestede* to Christ Church, Canterbury. *c*.1060 **Ketel** willed *Stistede* to Christ Church Canterbury (ECE 52,66) whose monks held *Stiesteda* in 1086,DB: stithe sted, nettle place. *(Fig. 61)*.

STOCK

In 1234,FF, Simon Blund rented 5 perches x 5 perches of land in *Herwardestoc* from Henry le Ture for 1d yearly: stocc, Hereweard's tree-stump.

STONDON MASSEY

Standon de Marcy, 1238,SR: stan dun, stony hill; Marci family, 13th cent, whose family originally came from Marcy, La Manche; Ralph de Marci held Kelvedon Hatch in 1086,DB. *(Fig. 6)*.

STORT RIVER

Stour, 1576,Saxton map: (Celtic) stur-, strong; *Stort*, 1586,Camden, back-formed from Bishops Stortford.

60. The bridge over Mar Dyke at Stifford and 17th-century Ford Place, c. 1871.

61. Stisted Hall in 1832.

STOUR RIVER

Stur muthan, ASC. In 884, King **Alfred**'s naval force opposed 16 Viking ships at the mouth of the *Sture* (but see Grainge, 2005). *(Fig.* **62***).*

STOW MARIES

Geoffrey de Mandeville held *Fenne* in 1086,DB; *Stowe*, 1222,FF: fenn stow, (holy?) place in the fen. Robert de *Marisco* rented out 1 messuage and 80 acres of land in 1250,FF; in 1374,Cl, John Mareys held 2 knights fees due for £10 to Joan, wife of Humphrey de **Bohun**, earl of Essex.

STRATFORD

Between 1066-87,(ECE 81), **William I** confirmed the grant of a mill at *Strætforda* to Westminster monastery: stræt ford, ford (through River Lea) at the (Roman) paved road.

STRATFORD LANGTHORNE

langan thorn, 958,(CS 1037); *Strætforde*, 1067,BL. In 1199,FF, Joia wife of Luke and Walter fitz Turstan quitclaimed to the Abbot of Hamme all land called *Langethorn in Hamme*: langan thorn, hamme, long thorn-tree at river-bend. *(See Stratford).*

STRETHALL

St. Etheldreda's, Ely Abbey held *Strathala* in 1086,DB: stræt halh, corner of land by the Roman street (from Great Chesterford to Brent Pelham).

STROOD CHANNEL, MERSEA

Fleote, 1046,(Hart,1992,p505); *Strodesflete*, 1455,AD: strod fleot, marsh-overgrown-with-brushwood inlet.

STURMER

Leofsunu spoke "No one round *Sturmere* will accuse me of deserting my master", Battle of Maldon poem, *c.*993. Tihel the Breton held *Sturmere*, 1086,DB: mere, pool by River Stour. See **Brihtnoth**. *See Coupals Farm.*

62. The River Stour near Dedham.

SUTTON
Swein held *Suttuna* in 1086,DB: tun, south settlement.

TAKELEY
St Valéry Abbey held *Tacheleia* in 1086,DB: Tæcca's or tagga leah, young sheep clearing.

TENDRING
Bishop of London, held *Tenderinga* in 1086,DB: ?people of Tønder, South Jutland or Tundern, Lower Saxony.

TERLING
c.1025,LE, Godgifu, widow of an ealdorman, willed *Terlinges* to Ely monastery. In 1086,DB, one hide was held by Jocelyn Lorimer from King **William** but was claimed by Ely: ingas, Tyrhtel's peoples' settlement.

TEY, GREAT, LITTLE
Teia was held by Count **Eustace** in 1086,DB: teag, enclosure.

TEY, MARKS
c.950,(CS1012), ealdorman **Ælfgar** willed *Tygan als Tigan* to his younger daughter. **Geoffrey** de Mandeville held *Teia* in 1086,DB: teag, enclosure. Adolf de **Merck** (near Calais), 1267,M.

THAMES RIVER
Tamesis, 51BC, Caesar: (pre-Celtic?) ta-, flowing. Compare Rivers Tamar, Thame (Oxon), and Tame (Staffs).

THAXTED
c.1000,LE, a venerable lady Ætheliva granted *Thacstede* to Ely monastery. Richard son of Count Gilbert held *Tachesteda* in 1086,DB: thæc stede, place where thatch-reeds grow (near headwaters of River Chelmer). *See Boyton and Yardley.*

THEYDON BOIS
Teidanam de Bosco, 1235,Ass: thæc dene, thatch-reed valley. In 1240,FF, Hugh de *Bosco* claimed from John de Lamburn one messuage in Affebrug (Abridge) for ½ mark of silver. As all three Theydons have their southern boundaries on River Roding, *thecdene* seems to apply to this stretch of the valley.

THEYDON GARNON
Thayden Gernun, 1238,Subs: thæc dene, thatch-reed valley. In 1201,FF, Ralph *Gernon* (of the family of **Robert** Gernon, Domesday tenant of Theydon Mount) was granted by Ralph fitz Peter 81½ acres in Bradefeld and service of Richard Bataill for one pound (weight) of pepper at Christmas. *(See Theydon Bois).*

THEYDON MOUNT
Theyden de Monte, 1254,Val: thæc dene, thatch-reed valley; church is on a hill. *(See Theydon Bois).*

THORPE-LE-SOKEN
*Torp(eia), Torpf, c.*1200,Colch. Scandinavian thorp, settlement. *See Walton.*

THORRINGTON
Odo held *Torinduna* in 1086,DB: tun, Scandinavian + Old English Thuri's or thorn-tree settlement.

THUNDERSLEY
Swein held *Thunreslea* in 1086,DB: leah, Thunor's wood/clearing.

THURROCK
*c.*1040,(ECE 48), Wulfstan the 'wild' priest granted *Turroce* to Christ Church Canterbury. Bishop of London held (Little) *Turrucca*, 1086,DB. Count of Eu (Seine Inferièure) held (West) and William Peverel (Grays) *Turruc*, 1086,DB: thurroc, ship-bilge.

TILBURY
*c.*735, **Bede** reported that **Cedd** built his church at *Tilaburg c.*653. William de **Warenne** held *Tilibia;* **Swein** held (West) *Tilibia*, 1086,DB: Tila's or til, lowland + burh, fort. Tilbury is surrounded by Thames marshes; the site of the fort has not been identified. *(Fig. 63).*

TILBURY-BY-CLARE
Tihel the Breton (of Helléan, Morbihan, Fig. **6**) held *Tilibia* in 1086,DB: til burh, useful or Tila's fort. ('Lowland' does not seem particularly appropriate here. *See Tilbury*).

63. Tilbury Fort about 1680.

TILLINGHAM
*c.*610, (CS 8), King **Æthelbert** of Kent granted *Tillingeham* to **Mellitus**, first bishop of London, for St. Paul's church. St. Paul's held *Tillingham* in 1086,DB: ingas ham, settlement of Tilli's people.

TILTY
Henry of Ferrers held *Tileteia* in 1086,DB: teag, Tila's enclosure. *(Fig. 64).*

TIPTREE
Tipentrie, 12th cent,Bodl: treow, Tippa's tree. *(Fig. 65).*

TOLLESBURY
(King **Eadred** granted 4 hides at *Tollesfuntum* in 946 to a religious woman Æthelgifu (Kelly 517a)). Barking monastery and Count **Eustace** held *Tolesbia* in 1086,DB: burh, Toll's stronghold (site of which is not identified). *(See **Erkenwald**).*

TOLLESHUNT
*c.*998, St Paul's required 1 man for ship duty from *Tollesfuntan* and Dunmow (ECE 28): funta, Toll's spring. Limes and Bowstead Brooks lie west of the present village; there are several ponds e.g. Pond Farm.

64. Tilty Abbey, 1784.

TOLLESHUNT D'ARCY

Odo held *Toleshunta* in 1086,DB: See above. Henry D'Arcy was Mayor of London in 1337. The D'Arcy family originated from Arci in Normandy in 1066.

TOLLESHUNT KNIGHTS

Count **Eustace** held *Toleshunta* in 1086,DB: See above. William de Bru(m)ford and Agnes, tenants, granted one *knight's* fee in Tolleshunt to Ralph Gernon, 1209,FF.

65. Tiptree and Maldon sign-post.

TOLLESHUNT MAJOR

In 1253,FF, John de Boville acknowledged the right of Thomas abbot of Coggeshall for 8d in rent and a ¼ knight's fee in *Tolleshunte Mauger.* The Archbishop of Canterbury's man Mauger was the undertenant of Robert son of Corbucion in 1086,DB.

TOPPESFIELD

Count **Eustace** held *Topesfelda* in 1086,DB: Topp's or hill-top open country. A plateau above River Colne lies at 76m O.D.

TOTHAM, GREAT, LITTLE

*c.*950,(CS 1012), ealdorman **Ælfgar** willed *Totham* to **Brihtnoth** and his younger daughter **Ælfflæd**. *Totham* was held by Bishop of London in 1086,DB: tot ham, look-out settlement. This is one of the earliest references in Essex to a look-out point (*Essex Journal*, 2007, vol 42, p. 11).

TOWER HAMLETS

Borough formed in 1965 from Bethnal Green, Stepney and Poplar which had belonged to the estate of the Tower of London.

TWINSTEAD

Richard son of Count Gilbert held *Tumesteda* in 1086,DB: twig or twinn + stede, brushwood or double place.

UGLEY

*c.*1045,(ECE 60), Wulfwin, son of Abbot Alfwin, granted *Uggele* to Ramsey Abbey, Cambs. *Vgghelea* was held by **Aubrey** de Vere in 1086,DB: leah, Ucga's wood/clearing. *(Fig. 33). See Bollington.*

ULTING

Ralph Baynard held *Ultinga* in 1086,DB: settlers by River Ult (?Celtic river-name, possibly associated with Old European 'al', to flow). 'Ult' may be the early name for this stretch of River Chelmer.

66. Upminster church and 13th-century tower in 1793.

UPMINSTER
In 1062, (K813), Earl **Harold** granted *Upmynstre* for the foundation of Waltham Holy Cross whose canons held *Vpmonstra* in 1086,DB: minster on upland. *(Fig. 66).*

UTTLESFORD
Udelesforda, 1086,DB: Udel's ford (at Wendens Ambo, through river Cam).

VANGE
In 963,(CS 1101), King **Edgar** sold *Fengge* to thegn Ingeram who, the same year, sold *Fenge* to Dunstan, archbishop of Christ Church, Canterbury (CS 1102). **Odo** and **Ranulf** Peverel held *Phenge* in 1086,DB: fenn ge, marsh district. *(Fig. 2).*

VIRLEY
Robert Gernon held *Salcota* in 1086,DB: salt cottages. *Salcote Verly*, 1291,For. Robert de *Verli* was Domesday subtenant, 1086. *(See also Salcott).*

WAKERING
Swein held *Wachelinga* in 1086,DB: ingas, settlement of Wacor's people or wacor, look-out peoples' settlement.

WAKES COLNE
In 1227,FF, Saher son of Richard Priest was granted 2½ acres in *Colum Malot* by Richard prior of Colum for 12d yearly: Celtic or pre-Celtic river-name. **Robert Malet** was Domesday holder in 1086. *Coln Wake*, 1375: Margaret wife of Baldwin de *Wake* (who died in 1282) inherited the estate.

WALTHAM ABBEY, HOLY CROSS
In 1062, (K813), Earl **Harold** founded *Waltham* Holy Cross. Bishop of Durham held *Waltha'* in 1086,DB: weald ham, forest settlement. Canonici *sancte Crucis de Waltham* were landholders in 1086,DB: church where was kept **Tofig's** miraculous holy cross. Edward I set up a cross to mark his dead queen Eleanor's penultimate

67. The Palladian Wanstead House.

resting-place before reaching Charing Cross, London. The Abbey was established in 1177. *(Fig. 12).*

WALTHAM FOREST
The borough established in 1965, amalgamating Chingford, Leyton, *Waltham*stow and Epping (*Forest*).

WALTHAM, GREAT, LITTLE
Robert son of Corbucion held *Waldham* in 1086,DB: weald ham, wood settlement. *(See back cover).*

WALTHAMSTOW
Between 1066-87,WDB, **William I** confirmed the grant of land at *Wilcumstowe* to Westminster monastery. Countess **Judith** held *Wilcumestou* in 1086,DB: Wilcume's or welcoming (holy) place.

WALTON-LE-SOKEN
*c.*970, (CS 737), King **Æthelstan** confirmed the grant of *Eduluesnæsa* (which probably included Thorpe and Kirby, *qv*) to St. Paul's: næss, Eadwulf's promontory. Socn, jurisdiction.

WALTON-ON-THE-NAZE
Walentonie, 11th cent.,StP: weala tun, Britons' or serfs' farm : næss, promontory.

WANSTEAD
In 1065,(K824), King **Edward** confirmed the grant of *Wænstede* by Ælfric to Westminster Abbey. St. Paul's held *Wenesteda* in 1086,DB: wæn or wenn stede, waggon or mound place. In 1715 a mosaic floor of a Roman villa was discovered in Wanstead Park. *(Fig. 67)*

WARLEY
*c.*1045,(ECE 46), **Leofgifu** willed 3 hides at *Werle* to Godwine her brother-in-law. Barking monastery held (Great) and Bishop of London held (Little) *Wareleia* in 1086,DB: waer leah, wood pledged (to Barking monastery). *(Fig. 68).*

WEELEY
*c.*1050, Eadgyva, with the consent of her husband, granted *Wilgelea alias Wigleya* in penance to St. Paul's (Kelly, 1243a). **Eudo** held *Wileia*, 1086,DB. William II confirmed Eudo's grant of *Wilega* to St. John's Abbey Colchester, *c.*1100 (ECE 119): wilig leah, willow wood/ clearing.

68. The 16th-century Hall at Little Warley in 1833.

WENDENS AMBO, WENDONS LOFTS

William of **Warenne** held *Wendena* in 1086,DB: wende denu, winding valley (of an easterly-flowing tributary of River Granta); (Latin) ambo = both; Great & Little Wenden were united in 1662; Robert Louhot had land at Lofts in 1236.

WENNINGTON

In 969,(CS 1264), King **Edgar** confirmed the grant of *Winintune* to Westminster Abbey. *c.*1044, (K870), King **Edward** confirmed its possession of the *burh* [fortified house] at *Winintune* as given by **Atsere** Swearte and his wife Ælfgyth. Westminster held *Wemtuna* in 1086,DB: tun, Wynna's settlement.

WEST BERGHOLT

Roger of Poitou held *Bercolt* in 1086,DB: beorg holt, hill-wood.

WETHERSFIELD

King **William** held *Witheresfelda* alias *Westrefelda* in 1086,DB: feld, Wihthere's open country.

WHITE COLNE

Colum Miblanc, 1141-63,Colne: Celtic or pre-Celtic river-name. Dimidius Blancus ("half-white"), a king's burgess, held one hide [roughly 120 acres], the manor of Ingledethorp in 1086,DB. *(Fig. 69).*

WHITE RODING

Roynges alba, white (probably lime-washed church), *c.*1121,WDB. In 1205,FF, William bishop of London granted to Alveva de Reines 1 cultura of land called *Witerudinge.* In 1239,FF, Hugh de Arden held 96 acres and a messuage from Roger de Cheyney in *Magna Roynges*, for 140 marks of silver: inga, Hrotha's peoples' settlement. *See Roding.*

69. White Colne church before and after 1870.

WICKFORD

*c.*980,(CS 1288), **Æthelflæd**, widow of King **Edmund**, willed *Wicforda* to her kinsman Sibriht. **Odo** and **Swein** held *Wicfort* in 1086,DB: wic ford, ford (through River Crouch) by the farm.

WICKHAM BISHOPS

Wicham was held by the Bishop of London in 1086,DB: wic ham, farm settlement.

WICKHAM ST. PAUL

*c.*970,(CS 737), King **Æthelstan** confirmed grant of *Wicham* to St. Paul's who required 2 men from *Hinawicun* (higna, community's wickham) and Tollandune for ship duty, *c.*998 (ECE 28). The canons held *Wicham* in 1086,DB: wic ham, farm settlement.

WICKEN BONHUNT

Gilbert son of Thorold held *Wicam* in 1086,DB; Banhunta, 1086,DB: wic ham, farm settlement; ban funta, dwellings at a spring where animal bones were sacrificed? or bon funta, helpful spring?

WID BROOK, WRITTLE

Wid River, 1848,White's map: probably back-formed from Widford, withig, willow. *(See also Blackmore).*

WIDDINGTON

St. Valéry Abbey held *Widituna* in 1086,DB: withegn tun, willow-tree farm. *(Fig. 6; Fig. 70).*

WIDFORD

Witford, 1216,Cl: withig ford, willow-tree ford (through Wid Brook).

WIGBOROUGH

King **Eadred** granted 17 hides at *Wicgan beorgun* to his faithful minister Ælfstan in 947 (Kelly, 522a). St. Marys, Barking held *Wigheberga* in 1086,DB: beorg, Wicga's hill/mound (perhaps on the hilltop above Layer Brook). *(See **Erkenwald**).*

WILLINGALE DOE, SPAIN

Ralph Baynard held *Willingehala*, and Count **Alan** held *Ulinghehala*, in 1086,DB: healh, nook of Willa's people, **Hugh D'Eu** (Seine Maritime), *c.*1170,M; Hervey de *Ispania* (Epaignes, Eure), was Domesday undertenant, 1086. *(Fig. 6).*

70. The 15th-century barn at Widdington.

71. Wivenhoe quay, River Colne.

WIMBISH

*c.*1036,LE, **Lustwine**, grandson of ealdorman **Brihtnoth** willed *Wimbisc* to Ely monastery. **Thurstan** son of Wine bequested an estate at *Wimbisc* to Christ Church Canterbury *c.*1042, (K788). **Ralph** Baynard held *Wimbeis* in 1086,DB: Wine's or winn, meadow + bysc, copse of bushes or beosic, reedy.

WITHAM

King **Edward** built a fort at *Wit ham, Witanham,* 912/3, ASC. **William I** held *Witham* in 1086,DB: wita ham, Wita's or councillors' settlement or wiht ham, settlement at the (Brain river) bend.

WIVENHOE

Wiunhov was held by **Robert** Gernon in 1086,DB: hoh, Wifa's ridge (above River Colne). *(Fig. 71).*

WIX

Hugh de Montfort and **Walter** the Deacon held *Wica* in 1086,DB: wic, farms.

WOODFORD

In 1062, (K813), Earl **Harold** gave *Wudeforde* to found Waltham Holy Cross whose canons held *Wdefort* in 1086,DB: wudu ford, ford (through Roding river) by the wood.

WOODHAM MORTIMER, WALTER, FERRERS

*c.*980,(CS 1288), **Æthelflæd** willed *Wudaham* to ealdorman **Brihtnoth** and her sister **Ælfflæd**. **Ranulf** Peverel held *Odeham* in 1086,DB: wudu ham, wood settlement; Henry II gave Woodham Mortimer to Robert de Mortimer's father, grandson of **Ranulf**. *c.*1000,(CS 1289), **Ælfflæd** willed *Wuduhamæs* to King **Æthelred's** mother Ælfthryth, then to Barking monastery. (But *Wdeham* was **Ralph** Baynard's in 1086,DB). Henry I gave Woodham Walter to Robert Fitzwalter (died 1134). **Henry** de Ferrers held *Vdeham*, 1086,DB; *c.*1087, (ECE 100), **William I** confirmed the grant of part of *Wodeham* by **Henry** of Ferrers to the Abbey of Bec, Eure. *(Fig.* **6***).*

WORMINGFORD

Robert Gernon held *Widemondefort* in 1086,DB: Withermund's ford (through River Stour at Mill Hill).

72. *Writtle Lodge in 1819.*

WRABNESS

Bury St. Edmund's Abbey held *Wrabenasa* in 1086,DB: Wrabba's or wrabb næsse, twisted promontory (above River Stour ending at Stone Point).

WRITTLE

King **William** held *Writela* in 1086,DB: writol, chattering, perhaps the name of Wid Brook, but *Writola burna* documented *c.*678 (CS 81) refers to Beam river, *qv*. *(Fig. 72).*

YELDHAM, GREAT, LITTLE

Richard son of Count Gilbert held (Great) *Geldham* in 1086,DB. Count **Eustace** held (Little) *Geldeham* in 1086,DB: gield ham, settlement liable for tax. *(Fig. 73). See also Spaynes Hall.*

73. *The ancient oak at Great Yeldham.*

Dictionary of Street, Road and Local names

National and Local Authorities are now responsible for giving names to new roads and streets. In the past as now, there seems to have been no general system. Names were acquired from usage, a prominent person (e.g. Duke Street, Chelmsford, from Richard Duk), the name of the field which the road crossed, a local building, landmark or topographical feature or some other characteristic which was sufficient to distinguish it. While many Authorities consult local history societies before naming streets, others seem to rely on Developers to do the research; this has led to 'misnames' such as "Hare Bridge Crescent" (in Heybridge, Ingatestone), but in fact the bridge was next to the Saxon enclosure, Old English *(ge)haeg* documented in the 14th century, and has nothing to do with hares!

The word 'road' derives from Old English 'rad', meaning 'a ride' but does not seem to have been applied to street-names until the 15th century. 'Street' (from English 'straet') implying a paved way is documented from the 8th century and probably first applied to Roman roads, thus 'Strethall', the corner of land by the paved road. Lanes (Old English 'lanu') were defined in the 13th century as thoroughfares wide enough for a cask of wine to be rolled along with one man at each side.

The oldest named route in Essex begins in Norfolk, crosses north Essex and continues to Wessex. It has been used by traders and travellers since prehistoric times, probably 4000 years or more. The *Icknield* Way is possibly a pre-Celtic name, for if the name is not related to the Iceni tribe of Norfolk, it seems to bear little resemblance to a Celtic, English, Scandinavian or Norman word. A suggestion that it might derive from Old English *Ica's hylte*, Ica's wood, seems improbable that so lengthy a route could be named from one person's wood. It may be a rare survival of a language used by Neolithic and Bronze Age people about which we know little or nothing. Though routes were built by the Roman military from London, through Chelmsford to Colchester, from Colchester to Great Dunmow towards Bishops Stortford, from Chelmsford to Long Melford (Suffolk), and many more some along paths long used by the Britons, the names they gave them (if indeed they did name them) are unknown. The Saxons called the Colchester to Braughing road *Stanstrete* (stone-paved street). The route north of Romford to Brentwood was *Here-straet* (army street).

More street-names are still in use from the 13th, 14th and 15th centuries. Maidenburgh Street, Colchester, in the 14th century was *Maydeneburghstraat* (street of the fort of the maidens); Burgess Well Road, Chelmsford, takes its name from a spring so named in 1341. Chipping Hill derives from Old English 'ciep', the early market place of Witham before it was moved to the main Chelmsford-Colchester Road by the Knights Templar *c*.1212. *Fisshamles* (in Barking) contains the element *sceamol*, a stall for goods. Vestiges of evidence of the Open Field System of farming in which 3 or 4 large fields were rotated between crops and pasture on a 3- or 4-yearly basis survive in Westfield (Laindon) and Eastfield Road (Walthamstow).

Agriculture and industry are represented in such street- and place-names as Smithies Farm (Sutton, blacksmithing), Tanhouse Lane (South Weald), where there was a tannery, Cornmill from the Cornmill Stream constructed by the monks and site of the mill at Waltham Abbey, Maltings Chase (Ingatestone) and Brick Kiln Lane (Horkesley Heath). Sadly, many fascinating names relating to associated activities such as *Cuckingstool End* (Saffron Walden) have now been 'sanitised' (to London Road), though it does survive at Newport where miscreants tied to a plank were tipped into Wicken Water. Field-names have given the names to many of the roads and streets that were built on them (Tenterfield Road).

The 18th and 19th centuries and the Industrial Revolution caused a rapid enlargement of many Essex towns and hence the name of streets. Wharf Road, Chelmsford locates the end of the Navigation canal from Heybridge, Maldon, and Factory Lane, Halstead is self-explanatory. In the 20th century increasingly popular became the use of View, Mead, Rise, Grove, Avenue, Gardens in towns perhaps as a reminder of a more rural environment.

Attempts to classify street-names along similar lines to place-names have resulted in subdivisions into Field/water names, Directional names, Religious, Trade, Buildings (including inns), Bridges, Personal, Commemorative and Thematic names (Room, 1992), Descriptive, Destination (Chelmsford Road), Possessive (Mildmay Road), Topical and Advertising names (Bebbington, 1972), and so on. Thus the Descriptive names include names as Cromwell Street; topical names include Alexandra Court, Cockaynes an early example of advertisement, though then perhaps not mercenary. This assortment still leaves some which seem to defy classification, but the classification is perhaps useful in trying to put some order into the melée.

Some idea of when the street or road was named (and therefore possibly established) can often be gained from the name itself, thus Waterloo Road (Brentwood), early 19th century, and Churchills Mews (Woodford Green) (whose MP he was) and Tensing Gardens (Billericay) from the mid 20th century.

This selection of street, road and local names here illustrates something of the people historically linked to Essex. In the main only those personages who had a significant association with Essex are included. For reasons of space, others of national significance are, in the main, excluded. People's names which appear in the chapter 'Who's Who' are in **bold type**.

ABBEY LANE, Coggeshall

Coggeshall Abbey (founded by Queen **Matilda** in 1142) of the Savignac Order to merge with the Cistercian in 1147. It was dissolved by Henry VIII in 1536.

ABBS CROSS GARDENS, Hornchurch

Aspiscrosse, 1514,Rental; William *Aps*, in late 12th century, was witness to a claim of a house and curtilage. *c.*1285,FF, Edward le *Aps* had a house on the Hornchurch-London road.

ABELS ROAD, Halstead

Abel de St. Martin was granted a market in 1255 by Hugh de Vere, earl of Oxford, (probably at Chipping Hill, Head Street).

ADAMS COURT, Halstead

Thomas *Adams* and his son Edgar in 1876 founded and developed Adams' Brewery.

AETHERIC ROAD, Braintree

The Saxon lord **Æthelric** and his wife Leofwyn gave *Boccinge* to the monks of Christ Church Canterbury for the salvation of his soul *c.*998, with the consent of King **Æthelred** (ECE 30).

AFFLECK ROAD, Colchester

Sir Edmund *Affleck* of Fingringhoe Hall (died 1788), MP for Colchester 1782, Admiral 1787.

AINGERS GREEN, Great Bentley

Angiers tenement was rented from Samuel Partridge,1708,ERxxiii,87: with Michael *Aunger*, 1319,SR.

AIRFIELD WAY, Hornchurch

Hornchurch Airfield was built at Sutton Farm in 1915. *See Squadrons Approach.*

ALDERFORD STREET, Sible Hedingham

Richard son of Count Gilbert held *Alreforda* in 1086,DB. Aluric of Alderford held part before 1066, DB. See Hedingham: alor ford, ford by alder-trees (through river Colne).

ALDERTON HALL LANE, Loughton

Ælwartone, 1062, (K813): tun, **Æthelward's** settlement? *(Fig. 14).*

ALEXANDRA COURT, Clifftown, Southend

Princess *Alexandra*, future wife of Edward VII, gave her name to the Alexandra Yacht Club, established 1883.

ALGERS ROAD, Loughton

Geoffrey *Algor* paid 3/8¾d. tax in Loketon in 1319/20,Loughton.

ALIBON GARDENS, Dagenham
Sir Richard *Alibon* (1621-88), born at Dagenham, judge of the Kings Bench.

ALLECTUS WAY, Witham
Allectus assassinated his associate Carausius to become usurper ruler of Britain 293-296 AD. On site of a Romano-British temple.

ALLEYNE WAY, Clacton-on-Sea
Edward *Alleyne*, 1566-1626, actor. He built the Fortune Theatre, Cripplegate and founded Dulwich College.

ALTHAM GROVE, Latton
James *Altham* of Latton, esq., died 1582/3, brother of Thomas, London merchant, had a house in Cheapside, London which he left to his son John.

ALSA STREET, Stansted Mountfitchet
Alsiehey, 1268,Ipm: (ge)haeg, Aelfsige's woodland enclosure. *See also Elsenham.*

AMBERDEN, Debden
*c.*1050, (K907), King **Edward** confirmed the grant of *Amerdene* to Ely monastery. **Ranulf Peverel** held *Amb'dana* in 1086, but Ely monastery claimed it: amore denu, yellow-hammer valley.

AMBRESBURY BANKS, Epping
Castrum de Eppynghatthe, 1299,WmC; *Amesbury*, 1670,M: ?burh, Ambr's fort.

AMPERS WICK, St. Osyth
*c.*1120, Bishop of London granted *Ampners wic* to the canons of St. Paul's; *Anmerswike*, 1539;TEAS,os,5,p5: Old French aumosnerie + wic, almoner's farm.

ANSGAR ROAD, Saffron Walden
The Saxon *Ansgar* held the manor of Walden before 1066,DB; it contained 66 villagers as heads-of-household. It was confiscated by **Geoffrey** de Mandeville before 1086.

ASCHAM DRIVE, Walthamstow
Roger *Ascham* (1515-68) tutored Elizabeth I while she was still princess, coming to live at Walthamstow.

ASHWELL HALL, Finchingfield
Walter *Cook* (possibly brother of **Richard** son of Count Gilbert) held *Assuuella* in 1086,DB: aesc wielle, ash-tree spring. *See Finchingfield.*

ASTRA CLOSE, Hornchurch
The RAF motto '*Per ardua ad astra*' remembers 2nd World War Hornchurch Airfield. *See Airfield Way.*

ATHELSTAN GARDENS, Wickford
Æthelstan, Saxon king of England, 924-939, grandson of King **Alfred**, opposed the Viking invasions. Buried at Malmesbury Abbey, Wiltshire.

ATLAS COURT, Earls Colne
Atlas Ironworks were built here from *c.*1866 by Reuben Hunt.

AUDLEY END, Gestingthorpe
Othelvesho, 1309,StJohn: Othulf's hoh, spur of land.

AUDLEY END, ROAD, Saffron Walden
Richard Turner, plaintiff, Thomas and Joan Garrold, defendants, 1 messuage in *Audleyend*, 1555,FF: Thomas, Baron *Audley* of Walden (1488-1544), Lord Chancellor, sanctioned the divorce from Catherine of Aragon of Henry VIII who granted him Walden Abbey.

AUGUSTINE WAY, Bicknacre
Augustine, monk sent by Pope Gregory to convert the pagan English AD 597. Baptised Kings **Æthelbert** of Kent and Sabert, king of the East Saxons, established bishoprics at Canterbury, Rochester and London, appointing **Mellitus** first bishop in Essex. Died 604. *See also* ***Sledd****, king of the East Saxons.*

AUGUSTUS WAY, Witham
Adopted by Julius Caesar as his son. Called Octavian. First Roman emperor AD 31-45.

AYLOFF'S WALK, Hornchurch
William *Ayloffe* (died 1585), judge, of Hornchurch. His son, William, was made a baronet in 1612. Sir Joseph, 6th baronet, (1709-81), FRS, was Vice-president of the Society of Antiquaries.

BACON END, Great Canfield
Beacon End, 1768,M. Site of a fire beacon at 85m OD. (*Essex Journal*, 2007, vol 42, p. 11)

BADER WAY, Hornchurch
Sir Douglas *Bader* (1910-82), 2nd World War RAF pilot, was based at Hornchurch Airfield.

BADDOW ROAD, Great Baddow
Richard de *Badew* (*c.*1320), born at Great Baddow, Chancellor of Cambridge University, founder of University (later Clare) College.

BALAAM STREET, West Ham
Balostret, 1371,Misc: associated with Walter *Balame*, 16th cent,MnAt.

74. Reconstruction of Balkerne Roman gate, Colchester.

BALDWINS HILL, Loughton
George *Baldwin* had two hearths in his house in 1666,Loughton.

BAKERS LANE, Ingatestone
Beast Market, *c.*1770,map (place of sale of livestock).

BALKERNE GATE, Colchester
Burghgate, 1346,Colch: burh, fortified; balca, boundary/beam. *(Fig. 74).*

BALL ALLEY, Colchester
John *Ball*, dissident preacher, took refuge in Bocking from the Bishop of London. He became chaplain in St. James' parish Colchester. Siding with the rebels of the Peasants' Revolt, he was hanged at St. Albans in 1381. Another John Ball (died 1394) was rector of the same parish in Colchester.

BALLARDS END, Great Stambridge
Ballardsgore, 1368,Harl: gara, point of land. Associated with *Ballard* of Rochford and Thomas atte Gore, 1319,SR.

BAMBER'S GREEN, Takeley
Benberys, 1496,Takeley: William *Bennebury* and John Bourke quitclaimed 162 acres of land, 14 acres of meadow, 6 acres of pasture and 4 acres of wood in 1422,FF.

BANCROFT AVENUE, Woodford
Francis *Bancroft*, draper, left money to the Drapers' Company for almshouses and a school in 1737. The latter moved to Woodford in 1886 from Mile End Road.

BANK STREET, Braintree
A boundary bank and ditch to the medieval town was present in the early 15th century.

BANNISTER GREEN, Felsted
Bernestey', 1235,Ass: teag, Beorn's enclosure? *See Felsted.*

BANYARDWAY, Rochford
James *Banyard* in 1838 founded the chapel of The Peculiar People whose belief was the certainty of Divine Salvation and, initially, a rejection of medical intervention.

BARING-GOULD COTTAGES, Mersea
Sabine *Baring-Gould* (1834-1924) rector of East Mersea, author of *The Broom Squire* and the hymn *Onward Christian Soldiers.*

BARRACK STREET, Bradfield
Site of the house for soldiers during the Napoleonic Wars.

BARKSTEAD ROAD, Colchester
Col. John *Barkstead*, MP for Colchester 1654, signed the death warrant of Charles I. He was executed in 1662.

BARLEYLANDS, Great Burstead
Barley Lands, 1777,C&A; John *Berley*, parson, of Little Ilford, and others, claimed 1 messuage, 240 acres of land, 15 acres of meadow and 90 acres of pasture in 1335,FF.

BARLEY LANE, Ilford
Named either from Dorothy *Barlee*, last abbess of Barking Abbey, 1539, or from Barley Marsh.

BARN HALL, Tolleshunt Knights
Ralph Baynard held *Borooldituna* in 1086,DB: Beornweald's tun, settlement.

75. Dr. Thomas Barnardo.

BARNADO DRIVE, Barking
Dr. Thomas *Barnado* (1845-1905), born in Dublin, founded 'homes' at Stepney in 1870 and Barkingside for destitute children. *(Fig. 75).*

BARRACK SQUARE, Moulsham
Barracks of West Essex Militia until 1873, then W.G. Wenley's furniture warehouse.

BARRICKS, High Easter
Odo held *Berewic* in 1086,DB: bær wic, barley farm. *See also Easter.*

BARRINGTONS, Chigwell
Robert Gernon held *Cingheuuella* in 1086,DB. Nicholas de *Barenton* (Hatfield Broad Oak) held it in 1274,Cl. *See Chigwell.*

BARRINGTON ROAD, Colchester
Sir Thomas *Barrington* (died 1644), a parliamentarian, was MP for Essex and Colchester.

BARTLOW END
In1247,FF, John de Wangeford held 2½ acres in *Berkelawe* from Richard and Alina de Kirkeby for one sore sparrowhawk: beorca hlaw, mound with birch-trees, possibly referring to the Romano-British burial mounds, originally 8 in number. *(Fig. 3).*

BARROW HALL, Little Wakering
Before 1066, **Finn** the Dane held *Berreuuera*, then **Richard** son of Count Gilbert in 1086,DB: beorg, barrow/hill. A tumulus is shown on the 6" OS map at TQ921880.

BARSTABLE, Basildon
Odo bishop of Bayeux held *Berdestestapla* in 1086,DB: berde, meaning uncertain, possibly bearded; stapol, pillar.

BARWELL WAY, Witham
Henry *Barwell*, clothier, built 'The Grove', demolished in 1932.

BASSETTS, Woodham
Walter Bishop of London held *Mildemet* in 1086,DB; middle maed, meadow (between R. Chelmer and woodland to the south); Richard *Basset*, 1304,Ipm.

BATA AVENUE, Tilbury
Tomas *Bata*, a Czech, founded the Bata Shoe Co in 1933, providing social housing, entertainment and sports facilities for his employees.

BATE-DUDLEY DRIVE, Bradwell-on-Sea
Sir Henry *Bate Dudley* (1745-1824), JP, clergyman, drained the Bradwell marshes and campaigned for improvements to the roads, edited the *Morning Post* and was imprisoned for libel.

BATTLESBRIDGE, Rawreth
Batailesbregge, 1351,Ass (over River Crouch). The *Bataille* family were of East Donyland; Reginald *Bataille*, 1327,SR.

BAYMANS WOOD, Brentwood
Percy *Bayman*, master of Brentwood School, gave land for the District Hospital in 1933.

BAYNING HOUSE, Little Bentley
Paul *Bayning's* (died 1629) granddaughter married the 20th Earl of Oxford, Aubrey de Vere.

BAYTHORNE END, Birdbrook
*c.*950, (CS 1012), ealdorman **Ælfgar** willed *Babingthirne* to his daughters **Æthelflæd** and **Ælfflæd** who *c.*1000 willed *Babbingthyrnan* to

Barking monastery (CS 1289). **Ranulf brother of Ilger** held *Babit'na* in 1086,DB: thyrne, Babba's thorn-bush.

BEAL ROAD, Ilford

Edmund *Beal*, 1843-1929, first Chairman of Ilford Urban District Council in 1895.

BEAUCHAMPS CLOSE, Chelmsford

Robert de *Beauchamp* fought at Hastings, 1066. From his brother Henry, first earl of Warwick, descended the long-surviving, powerful Beauchamp then Neville families.

BEAUCHAMPS DRIVE, BELCHAMPS ROAD, Wickford

Associated with John de *Bello Campo* in 1306,Cl. See above and Beauchamp Roding.

BEAUMONT HILL, Great Dunmow

Sir George *Beaumont* (1753-1827) of Clock House, painter, helped to found the National Gallery.

BECHE ROAD, Colchester

Thomas *Beche*, abbot of St. John's, refused to submit to Henry VIII's will of Dissolution of the abbey and was executed.

BECK ROAD, Saffron Walden

Sir Arthur Tyrrell *Beck*, Liberal MP for Saffron Walden 1910-1925.

BECKINGHAM, Tolleshunt Major

associated with Stephen *Beckinghame*, 1543,MnAt, from Beckingham, Lincs or Northants.

BECKNEY, Ashingdon

Bacheneia was held by **Odo** in 1086,DB: Becca's low-lying land. *See also Ashingdon.*

BECKTON, East Ham

Named after the Governor of the Gas, Light & Coke Co., S.A. *Beck*, 1869.

BEELEIGH, Maldon

Byleg, 1225,Pat: beo leah, bee clearing. Honey was used as a sweetener, candle wax and for making mead.

BEDELLS AVENUE, Black Notley

William *Bedell* (1571-1642), born at Black Notley, was Provost of Trinity College Dublin and became in 1629 bishop of Kilmore and Ardagh in Ireland.

BEDLAR'S GREEN, Great Hallingbury

Bedewell, 1229,Cl: byde wielle, stream in a hollow. This seems to be the stream which runs westwards into the River Stort.

BELSTEAD HALL, Chelmsford

William de **Warenne** held *Belesteda* in 1086,DB: belhus stede, place of a bell-house.

BENDLOWES ROAD, Great Bardfield

William *Bendlowes* (1516-84), sergeant-at-law, bought Place Farm *c.*1564. His great-grandson Edward of Finchingfield (died 1676), poet, was a benefactor of St. John's College, Cambridge. *See Berwick Lane.*

BENDYSH HALL, Radwinter

Count **Eustace** held *Benedisc* in 1086: bene edisc, bean enclosure.

BENSUSAN COURT, Colchester

Samuel *Bensusan* (1872-1958) recorded the now-bygone life in the Essex marshes and rural countryside.

BENTFIELD END, Stansted Mountfitchet

Robert Gernon held *Benedfelda* in 1086,DB: beonet feld, bent-grass open country.

BENTON CLOSE, HALL, Witham

Bury St. Edmunds monastery held *Breddinchou* in 1086,DB: ?Bridd's hoh, promontory. *See Witham.*

BENYON PATH, South Ockendon

From Mr. H. Benyon's estate was bought the village hall in 1964. The family had been landholders here since 1758.

BERIFFE PLACE, Brightlingsea

John *Beryf* (died 1496), William *Beriff* of 'Jacobes' (died 1578) have brasses in the north chapel of All Saints church which contains a Norman door . The family were merchants.

BERWICK BERNERS, Abbess Roding

Eudo the Steward held *Roinges* in 1086,DB. Beatricia (once wife of William de *Berners*) quitclaimed as dowry 1/3 of *Berewic* manor, 1220,FF: bær wic, barley farm. *See Roding.*

BERWICK LANE, Stanford Rivers

In 1563,FF, Thomas Patteshale and Thomas Foster, gents, granted a messuage and the manor of *Berwykes* to William Bendlowes, serjeant-at-law, and Richard Elliot, gent. Old English bere wic, barley farm, later an outlying farm. *See Bendlowes Road.*

BERWICK, Rainham

Robert Gernon held *Renaham* in 1086,DB. *Berwyk*, 1361: bær wick, barley farm. *See Rainham.*

BICKNACRE, Woodham Ferrers
Bikenacher, 1186,P: æcer, Bicca's cultivated land.
BIGGIN, Chadwell
Byggynge, 16th cent,MnAt: Middle English *bigging*, an outbuilding. This was one of the four manors formed by the Bishop of London in 12th century.
BIGGIN LANE, Tilbury
Middle English (formerly a Scandinavian loan-word) *bigging*, an outhouse.
BIGODS LANE (Alfriston), Great Dunmow
Geoffrey de Mandeville held *Alferestuna* in 1086,DB: tun, Ælfhere's settlement, Bartholomew le *Bigod*, tenant, 1235,Ass.
See Dunmow.
BIRCH HALL, Kirby le Soken
Count **Eustace** held *Birichou* in 1086,DB: bierce hoh, birch-tree promontory.
BIRD GREEN, Clavering
Thomas and Dorothy *Byrde* quitclaimed to Henry and Mary Ward 1 messuage, 80 acres of land, 12 acres pasture in 1561,FF.
BISHOPS HALL ROAD, Chelmsford
William, *Bishop* of London, 1051-1075, had a 600 acre manor with a mill here. Successor Bishop Maurice (died 1107) was William I's chancellor, rebuilt St. Paul's and built the Great Bridge (now site of Stone Bridge at Chelmsford), diverting the London road from Writtle. *(See Lawford Lane)*.
BLACK CHAPEL, Great Waltham
Robert Reignold surrendered a croft against *Bleccheschapel* to John and Joan Wrongy, 1376,Ct. Aylwyn *Bleche* in 1246,FF, held land west of Langefeld. In 1376,Ct, John *Blecche* was farmer here. *See Great Waltham*.
BLAMSTERS CRESCENT, Halstead
William de *Albo Monasterio* quitclaimed to Alan de Crippinge the advowson of Haudstede church, in 1233,FF. John atte Nunnotehey to hold 4½ acres in Halstede from John de *Blaumuster* for rent of a clove of a gillyflower, in 1286,FF.
BLAMSTER'S HALL, Great Easton
William de **Warenne** held *Estanes* in 1086,DB; *Eystan Blaumost'*, 1285,Ass: William de *Blauminister*, 1280,Ipm, (Whitchurch, Shropshire).

BLIND KNIGHTS, Layer de la Haye
Blindeknygthes, 1364,Colch: land given by Richard I for blinded crusaders.
BLIND LANE, Little Burstead, Howe Green and W. Hanningfield
A cul-de-sac.
BLUEBRIDGE, Halstead
Aissesford bridge, 1255,Colne: Aesc's or ash-tree ford. *Blew Bridge*, 1722, when John Morley, butcher, painted it blue. *See Morley Road*.
BLUNT'S HALL, Witham
Count **Eustace** and **Ranulf Peverel** held *Blundeshala* in 1086,DB: healh, Blunt's corner.
See Witham.
BLUNTS WALL ROAD, Billericay
Reginald *Blund*, of the manor house here in 13th century, sided with Simon de Montfort and subsequently had his lands confiscated to the monks of Thoby Priory, Mountnessing.
BOADICEA WAY, Colchester
Queen of the Iceni tribe (of Norfolk), and widow of King Prasutagus, she raised a rebellion against the Roman administration in 60/61 AD, sacking Colchester, London and St. Albans before being defeated in the Midlands. *(Fig. 76)*.
BOBBINGWORTH, Bovinger
Ranulf brother of Ilger held *Bubingeorda* in 1086,DB: worth, Bubba's people's enclosure.
BOBLOW, Helion Bumpstead
Boblowe, 12th century,StJohn: Bubba's (burial?) mound.

76. Statue of Boudicca by Thomas Thornycroft, Westminster Bridge, unveiled 1902.

BOCKING HALL, West Mersea
*c.*998, **Æthelric** and his wife Leofwyn granted *Meresege* to Christ Church Canterbury (ECE 30). The monks held 2 hides at *Meresai* in 1086,DB: eg, pool by an island.

BOHUN CLOSE, Great Leighs
Humphrey de **Bohun**, father of the 1st earl of Essex (died 1274), married Maud, heiress of the de Mandevilles who had acquired Leighs after the Norman Conquest, 1066.

BOIS FIELD TERRACE, Halstead
Boyes, a messuage, 1428,FA: John de *Bosco*, in 1227,FF, represented Richard Pirot who quitclaimed to Jordan le Brun the advowson of Hakewell church. (Latin boscum, French bois=wood).

BOLFORD STREET, Thaxted
Boldfordstreate, 1597,Rental: bula ford, bull-ford.

BOLEYN WAY, Boreham
Ann *Boleyn* celebrated the birth of her daughter Elizabeth at New Hall in 1533. The estate had belonged to her father Thomas before Henry VIII acquired it in 1517.

BOLINGBROKE CLOSE, Great Leighs
Through his marriage to the **Bohun** heiress of Essex, Henry IV (*c.*1366-1413), born at Bolingbroke, Lincolnshire, held Great and Little Leighs.

BOLLINGTON HALL, Ugley
Robert Gernon held *Bolituna* in 1086,DB: Bola's ing tun, settlement place.

BOOSE'S GREEN, Colne Engaine
Bouser, 1493,Ipm: associated with Robert de *Burser*, 1285,Ass.

BOOTHBY COURT, Chingford
Thomas *Boothby* (died 1622), merchant, bought the Manor of Chingford; his son Thomas was created baronet in 1660.

BORDERS LANE, Loughton
William the *bordar* (free man who held his dwelling from the lord of the manor in medieval times) witnessed a land grant in Theydon, ADi.

BOSWORTH CRESCENT, Romford
Joseph *Bosworth* founded the Romford charity school in 1710.

BOURCHIER'S HALL, Aldham (alias Little Fordham)
Richard son of Count Gilbert held *Sordeha* (recte *Fordeha*) in 1086,DB: ham(m), settlement or corner of land by a ford. *c.*1311,M, John *Boussier* acquired it with a watermill and messuage and 180 acres of arable land.

BOURCHIER WAY, Halstead
John de *Bourchier*, judge, (died 1328, buried in Halstead church) held half the manor of Stanstead Hall. His son Robert, 1st Lord Bourchier, Lord Chancellor, (died 1349), fought at Crécy.

BOVILLS UPLANDS, Mayland
Bovyles, 1426,Ipm: associated with Thomas de *Bouyle* of Bradwell-on-Sea, 1285,Ipm.

BOVILLS WAY, Little Clacton
Bevyles,1394,MnAt; Stephen de *Boville* held Bovill's Hall before 1272,Colch.

BOXMILL LANE, Halstead
Site of *Boxe Mill*, 1555,Rental, on River Colne.

BOYTON HALL, Finchingfield
Richard son of Count Gilbert held *Boituna* in 1086,DB: boia tun, a servant's or Boia's settlement.

BRADWELL, St. Peter's Chapel
Othona, 5th cent; **Bede** in 735 reported that **Cedd** built a church at *Ythancæstir c.*653: ceaster, ?Otto's fortress. In 1068, **William I** granted *Wallam* to **St. Valéry** Abbey, France. The Abbey held *Effecestra* in 1086,DB. *Wallam* refers to the Roman fort on which the church was built. *See Bradwell-on-Sea. (Fig. 6; Fig. 7).*

BRAN END, Stebbing
Brandonande, 1422,Rental: brende dun ende, burned hill district.

BRANFILL ROAD, Upminster
Andrew *Branfill*, sea-captain, bought Upminster Hall in 1685, made his fortune in gold.

BREACH LANE, Dagenham
An exceptional Thames tide burst a sluice-drain on 29th October 1707 flooding 1000 acres of Dagenham Levels west of Beam river. The breach was finally closed in 1720 at a cost of over £40,472.

BRENT HALL, Ugley
Swein annexed *Bertuna* in 1086,DB: bere tun, an outlying grange barley farm (?of Bollington Hall).

BRENTHALL, Harlow
Ranulf brother of Ilger held *Herlaua* in 1086,DB, brende, burnt. *See Harlow.*

BRETTS, Aveley
Helituna was held by **Swein** in 1086,DB. John le *Brett*, 1206,Abbr.

BROCKWELL LANE, Kelvedon
Brokhole, 1289,Survey: Old English brocc hol, badgers' sett.

BROOK STREET, Brentwood
Brokstret, 1479,Will: Middle English broc, stream. *Sideburnbroc*, 1198,Cur: sid, burh, broc, stream by a wide earthwork, may refer to the Iron Age hillfort in Weald Park but a stream does not flow from here to Brook Street now. A stream does flow south beside Brook Street from Brentwood town. The street is the Roman road between Romford and Brentwood.

BROOMSHAWBURY, Hatfield Broad Oak
Bronesho, 1299,Ipm: hoh burh, Brun's manor on a promontory.

BROWNE CLOSE, Brentwood
Sir Anthony *Browne* (1510-1567) of South Weald, Chief Justice of Common Pleas, founded Brentwood School in 1557.

BRUFF DRIVE, Walton on Naze
Peter Schluyer *Bruff*, civil engineer, extended the Tendring Hundred Railway to Walton, 1867.

BUILDING END, Chrishall
Agnes de Sumari claimed from Henry de Pinkeni 13 acres next to the wood of Reinbald in *Billinden* in 1200,FF. Billa's ing denu, place in the valley.

BULFORD MILL LANE, Cressing
Bolefen mille, 14th cent.,Survey: bulls' marshland (by the ford through River Brain) probably the site of the mill mentioned in Domesday Book. *(Fig. 77).*

BUMPSTEAD, Aveley
Odo held *Aluitheleia* in 1086,DB: Ælfgyth's leah, wood/clearing. ? bune stede, reedy place. *See Aveley.*

BUNYAN ROAD, Braintree
John *Bunyan* (1628-88), friend of John Taylor, is reputed to have preached in Braintree at Edward English's barn *(see Taylor Avenue).*

BURGESS WELL ROAD, Chelmsford
Named from a spring (now in the garden of 15 Fairfield Road), Burgeyswell, 1428, which supplied water to the 'Dominican' priory. John *Burgeys*, friary benefactor, 1281,Grieve.

77. Bulford Mill at Cressing, on the River Brain.

BURKITT'S LANE, Dedham
Rev. William *Burkitt* (1650-1703), vicar of Dedham 1692-1703, author.

BURNS CLOSE, Billericay
Until 1998 here was the Regional Burns Unit, St Andrews Hospital, *qv. (Fig. 78).*

BURGOYNE HATCH, Harlow
Montagu *Burgoyne* (1750-1836), bought Mark Hall *c.*1780. Verderer of Epping Forest, Whig MP.

BURTON BOWER, Stansted Mountfitchet
In 1315,FF, Edmund le Hunte claimed 2 messuages, 5 acres in Farnham of John and Agnes de *Burton*; paid one sore sparrowhawk. OE. bur, cottage.

BURY LANE, Epping
Leads to Epping Bury, the manor house of the priests of Waltham Abbey.

78. The Regional Burns Unit at Billericay, in 1994.

BUTLER CLOSE, Saffron Walden
Richard Austin *Butler*, MP for Walden 1929-65, Deputy Prime Minister, Baron and Master of Trinity College, Cambridge, 1965.

BUTT ROAD, Colchester
Butt Lane, 1748,M: named from the archery butts, or Middle English butte, boundary-land.

BUTTERCROSS LANE, Epping
The Butter Cross here was removed in 1781.

BUXEY SAND, east of Tillingham
Bokesey, *c.*1310,Takeley: Old Norse ey, Bucc's island (now inundated by the sea).

BUXTON ROAD, Coggeshall
Robert *Buxton* married Emma Paycocke in 16th century. Sir Thomas of Earls Colne (1786-1845) assisted his sister-in-law Elizabeth Fry (1780-1846) to improve the prison system.

BUXTON ROAD, Theydon Bois
Thomas Fowell *Buxton* (1837-1915), 3rd baronet, of Warlies, Epping, MP, Governor of Australia. His brother Edward of Knighton, Buckhurst Hill (1840-1924), JP, Alderman of Essex County Council, Verderer of Epping Forest, Liberal MP 1885. Gerald *Buxton* lived at Birch Hall, 1900.

BYRD MEAD, Stondon Massey
William *Byrd* (died 1623), composer of church music and madrigals, lived at old Stondon Place. He was organist to the Chapel Royal of Elizabeth I despite holding to the Catholic faith.

BYRTON, Stanway
*c.*1000,(CS 1289), **Ælfflæd**, widow of ealdorman **Brihtnoth**, willed *Byrætune* to King **Æthelred**. Roger Marshal held *Bertuna* in 1086,DB: bær tun, woodland swine-pasture farm (was on the borders of Lexden and Stanway).

CABBORNS CRESCENT, Stanford-le-Hope
Cabournes, 1470,ECP; John and Joan *Caborne* claimed 10 acres in Stanford 'in the Hope' from Robert and Beatrice Clobber for 10 marks, 1429,FF. *(Fig. 79).*

CAGE END, Hatfield Broad Oak
The site of the parish lock-up.

CAIDGE, Southminster
la Kage, 1323. Walter fitz Wauter of Wodeham quitclaimed manors of Cage and Crixhith to John Lewis and others for 200 marks in 1546,FF: Old French cage, enclosed land, also a lock-up.

79. The 15th-century Cabborns Manor Farm at Stanford-le-Hope.

CALCOTT CLOSE, Brentwood
John de Depeden and Joan to hold from Nicholas de *Caldecote*, vicar of Welde (South Weald), the manor of Laufare, 1345,FF: cold shelter. Calcott Hall lay west of the bypass.

CAMBRIDGE BROOK, Mount Bures
Kenborne, 1410,MnAt: cyna burna, cows' stream.

CAMPBELL AVENUE, Ilford
Dr. Sidney *Campbell*, 1910-74, born at Barkingside, became organist to Ely Cathedral and Windsor.

CAMPION ROAD, Colchester
Sir William *Campion*, Kentish royalist, was killed defending the town against Colonel Fairfax, 1648.

CAMPION WAY, Witham
George B. *Campion* (1796-1870), artist and painter, illustrated Newland Street in Wright's *History of Essex*.

CAMULODUNUM WAY, Colchester
The 'duno' or fort of the Celtic god of war *Camulos* was the the centre of the Trinovantes tribe who submitted to Julius Caesar in 54 BC. Archaeology has sited this at Gosbecks.

CANFIELD ROAD, Canfield
Benet *Canfield* (1563-1611), alias William Fitch, Capuchin friar, was imprisoned in the Tower, but later released to the Convent at Rouen, France. The place-name predates the friar by at least 500 years.

CANGLE, Halstead
Associated with Adam de *Gangel*, 1235,Ass: Middle English cangel, fenced enclosure.

CANN HALL, Wanstead
Leintuna was held by **Hugh** de Montfort in 1086,DB. *See Leyton*.

CANNHALL ROAD, Wanstead
Cann Halle manor, 1276,ADv, belonged to the canons of Holy Trinity, London.

CANNOCK HILL COURT, Colchester
Wicam Canonicorum, 1158,Colch: wic, dairy farm of the canons of St. John's Abbey.

CANONS AND PASSMORES, Great Parndon
Ranulf brother of Ilger held *Perenduna*, 1086,DB: his undertenant Roger's holding became Canons (of Beeleigh monastery); Alured's became Passmores, 1086,DB; Ralph *Passemer*, 1198,CurR. *See Parndon*.

CAPEL ROAD, Colchester
Lord Arthur *Capel* (1610-1649), baron of Hadham Herts, took part in defence of Colchester against Colonel Fairfax, 1648. He was executed at the Tower, but his son Arthur (1631-83) was created Earl of Essex on the Restoration of Charles II in 1661.

CAPEL CLOSE, Rayne
William *Capel*, Lord Mayor of London in 1503, lived at Rayne Hall. His son Sir Giles (died 1520) rebuilt it.

CARMEN STREET, Great Chesterford
associated with John son of William *Calman*, *c*.1470,ECP.

CASTLE STREET, Chipping Ongar
The street marks the southern wall of the bailey of the motte-and-bailey castle built before 1157, possibly by Richard de Lucy, the king's Justicar. The keep was demolished in 16th century.

CAVENDISH AVENUE, Colchester
Margaret *Cavendish* (1625-74) wife of Marquis of Newcastle and daughter of Thomas Lucas of St. John's, Colchester *(see Lucas Road)*. 'Unsuccessful' playwright, buried in Westminster Abbey.

CAVENDISH LANE, Pentlow
John de *Cavendish*, Chief Justice of the King's Bench, was beheaded in 1373 by Wat Tyler's followers during the Peasants' Revolt.

CAWKWELL CLOSE, Chelmer Village
Thomas *Cawkwell*, first clerk to Chelmsford Town Commissioners, *c*.1789.

CHALLIS LANE, Braintree
James *Challis* (1803-82) was Professor of Astronomy at Cambridge, and observed the planet Neptune 6 weeks before it was "discovered" by Dr. Galle in 1846.

CHAMPION CLOSE, Stanford le Hope
Richard *Champion* of Hassenbrook, held lands in Tilbury, Burstead, Horndon, in Reading and Suffolk. Buried in Stanford le Hope church, 1600.

CHANCELLOR AVENUE, Chelmsford
Frederick *Chancellor* (1825-1918), architect of Felsted School; mayor of Chelmsford. His son Frederick Wykeham, FRIBA, (1865-1945) restored St. Cedd's chapel at Bradwell, Leez Priory and Layer Marney towers *(Fig. 46, Fig, 47, Fig. 80)*.

80. *Frederick Wykeham Chancellor, FRIBA.*

CHAPMAN SANDS, Canvey
Chapmansand, 1402,Pat: Middle English *chapman*, merchant.

CHARDWELL (*alias* Ainsworth), Arkesden
William de **Warenne** held *Einesuurda* in 1086,DB: Ægen's worth, enclosure. *Cheldewell*, 1361: cald wielle, cold spring.

CHATHAM GREEN, Great Waltham
Geoffrey de Mandeville held *Cetham* in 1086,DB: (Celtic) cet + ham, wood settlement. See Waltham.

CHAURETH HALL, Broxted
Richard I of Clare (Suffolk) granted the tithes from *Chavreia* to monks of Bec Abbey from his man Warner, *c.*1080 (ECE 78). Warner, a man of **Richard** son of Count Gilbert, held 30 acres in *Ceauride* in 1086,DB: Ceawra's rith, brook. *See Broxted.*

CHIPPING HILL, Witham
Chippinghill, 1608: Old English cieping, market-place. *See Newland Street.*

CHISWICK HALL, Chrishall
Robert of Tosny (Eure) held *Ceseuuic* in 1086,DB: ciese wic, cheese farm.

CHRISTY AVENUE, Chelmsford
Miller *Christy* (1861-1928), Essex historian, co-founder of the Essex Field Club, lived at Chignall.

CHURCHILLS MEWS, Woodford Green
Winston Spencer *Churchill* 1874-1965, MP for Woodford 1945-64, Knight of the Garter 1953.

CINDERY ISLAND, Brightlingsea
Syndry, 1539. sundor eg, low-lying land cut off (from mainland).

CLARET HALL, Ashen
Count **Eustace** held *Clare* in 1086,DB: it lies opposite *Clara*, Clare, Suffolk, ?a stream name.

CLARKS ROAD, Ilford
Sir Allen *Clark*, 1898-1946, director of Plessey Aircraft Engineering Company at Ilford.

CLATTERFORD END, Fyfield
Claterfoot End, 1777,C&A: clater ford, loose stone or noisy ford.

CLAUDIUS WAY, Witham
Claudius, Roman emperor AD 41-54 who invaded Britain in AD 43 spending 16 days in the country set up his army's fort at Colchester. Mortally poisoned by his wife Agrippina.

CLAYS LANE, Loughton
In 1738, Richard *Clay*, draper, surrendered Plumtree Mead to William Hamilton of Lincoln's Inn. Richard Lomax *Clay* esq. resided at Golden Hill, 1774. *See Goldings Hill.*

CLIFF FARM, Mistley
Roger of Raismes (Nord) held *Cliva* in 1086,DB: clif, cliff. *See Mistley.*

COATES CLOSE, Heybridge
Richard *Coates* supervised the construction of the Chelmer and Blackwater Navigation under John Rennie, completed in 1797. *(Fig. 81).*

COCKAYNES, Alresford
Cokayne, 1279,FF; Benedicta de Cokefeld and Avice acknowledged the right of Ranulf Parson of Lalleford to hold 1 messuage, 226 acres etc. in Elmstede, *Cokayne*, Fratyngg etc, 1311,FF: a mythical paradise of delight (14th century poem).

81. Sandford Lock on the River Chelmer.

COLDHARBOUR LANE, Wennington
Coleherbert, 16th cent.,LP. 'Coldharbour' was usually used to denote a distant upgraded miserable dwelling. This Essex example, with one in High Ongar, is one of the earliest documented in England.
COLLEGE WOOD, Fryerning
Dorothy widow of Sir Nicholas Wadham (1532-1609) granted this to Wadham *College*, Oxford, in 1612. *See Wadham Chase.*
COLLIER ROW, Romford
Colyer Rowe, 1453,ADvi: Middle English, colier raew, charcoal-burners' row of houses.
COLLIER STREET, Hatfield Broad Oak
Coldwell, 13th cent,TEASi,82: cald wielle, cold stream/spring
COLLIERS SPRING, Navestock
Geoffrey le *Colyer*, 1319,SR: colere, charcoal-burner + spring, wood.
COLLINS CLOSE, Braintree
Samuel *Collins*, vicar of Braintree 1610-56, turned as much of a blind eye to the Dissenters in his parish as he could despite the pressure of his bishop Laud, later archbishop of Canterbury.
COLNEFORD HILL, Earls Colne
The corn mill was situated in Lower Holt Street, demolished in 1897.
COMYNS PLACE, Writtle
Sir John *Comyns* (1667-1740), judge, MP for Maldon, Chief Baron of the Exchequer, built Hylands House *c.*1730.
CONDUIT STREET, Chelmsford
A water conduit was erected in 1814, moved to High Street in 1852, then in 1940 to Tower Gardens, Rainsford End.
CONIES ROAD, Halstead
Associated with Conies Farm; Richard *Coney*, 1559.
CONSTANTINE ROAD, Witham
On his father's death, proclaimed Roman emperor AD 306 in York after fighting the invading Picts. Issued Edict of Milan in 313 allowing Christians freedom to worship openly.

82. Copperas collection from the beach at Harwich, 1730.

COOKES CLOSE, Romford
Sir Anthony *Cooke* of Guydyhall (Gidea Hall) 1504-1576, MP for Essex, father of Mildred wife of Lord William Cecil Burghley of Hatfield.

COOPERSALE LANE, Epping
To John Roger of *Cobershale,* land in Theydon Garnon, *c.*1450,TEASv,23: possibly personal name + Old English halh, corner of land. *(Fig. 17).*

COPPED HALI, Epping
Coppedehall, 1258,Wm: copped, pinnacled.

COPPERAS ROAD, Brightlingsea
Copperas: ferrous and copper sulphate used as black dye and cure for sheep-scab was collected in the 18th century along the northeast Essex coast. *(Fig. 82).*

COPPERMILL LANE, Walthamstow
Coppermill Stream and *Coppermill Farm,* 1875,OS.

CORBETS TEY, Upminster
Henry *Corvyn,* 13th cent,HPD; Richard and Joan Dryland quitclaimed to Richard Broke, serjeant-at-law, land in Upminster, *Corbynsty,* 1514,FF. In 1550,FF, John *Corbyn* and William Rotheley of London, goldsmith, granted to Ralph Josselyn, alderman, 1 messuage and land in Kent, and 16 acres land, 16 acres meadow, 4 acres pasture in Berkyng for 100 marks.: teag, enclosure.

CORNISH HALL, Finchingfield
Mascerel (Old French, 'the grinder') held *Nortuna* in 1086,DB: north tun, settlement. *Cornerthys,* 1381, associated with Richard of *Cornard,* near Sudbury, Suffolk. *See Finchingfield.*

CORYTON, Corringham
Kynochtown, the site of Kynoch's ammunition factory from 1896, was bought by (Sir) Clifford *Cory* (died 1941), in 1923 as an oil-refining depot, in 1953 Mobil Co. The village was demolished in the 1970s

COSTED MANOR ROAD, Brentwood
Costhall *alias* Costed Hall, 1537,M. *See Coxtie Green.*

COUPALS *alias* CHELVESTON, Sturmer
Celuestuna was held by **Richard** son of Count Gilbert and **Ralph** Baynard in 1086,DB: Ceolwulf's tun, settlement. William de Gerdel and Richard Brun quitclaimed to William de Grantcurt and Hugh and Ela *Curtpeil* ½ knights fee in Sturmer for 26 marks of silver in 1232,FF.

83. *Ftizwalters, Shenfield, the seat of T. Wright, 1818.*

COURAGE WALK, Hutton
The *Courage* family, brewers, including R.H. *Courage*, JP, benefactors to Brentwood and Shenfield, lived at Shenfield Place, built *c.*1690, then Fitzwalters. *(Fig. 83).*
COURTAULD CLOSE, Halstead
Samuel *Courtauld* acquired Townsford Mill in 1828 as a silk-weaving factory. *(Fig. 84).*
COURTAULD ROAD, Braintree
George *Courtauld* (1761-1823) moved the family silk textile business to Braintree in 1809. *c.*1854 his son Samuel purchased Gosfield Hall. The firm gained an international reputation.
COVAL LANE, Chelmsford
Coval Hall, residence of Robert Tindal, solicitor, 1790, named from Geoffrey *Colvyle*, 14th century. Before 1851, called *New Barrack Lane*, site of barracks 1796-1814.
COWBRIDGE FARM, Mountnessing
Cubrigea was held by **Ranulf brother of Ilger** in 1086,DB: cu brycg, cow bridge (over R. Wid).
COWDRAY CRESCENT, Colchester
Viscount *Cowdray*, MP for Colchester, purchased the castle for the Borough in 1920.
COXTIE GREEN, South Weald
Cocstede, 1177-84,Ch: cocc stede, cockerel place.
CRAMPHORN WALK, Chelmsford
T.J.D *Cramphorn*, corn merchant, occupied 17 High Street in 1787. The business expanded throughout the 19th century.

84. *Townsford Mill, Halstead.*

85. The Creffeild family of Colchester, by William Wellings, 1794.

CRANES FARM ROAD, Nevendon
Cranes, associated with Hugh le *Crane* before 1272,Colch.

CRAWLEY END, (-bury), Chrishall
Count **Eustace** held *Crauuelea* in 1086,DB: crawe leah, crow wood.

CRAXE'S GREEN, Birch
John *Crax* quitclaimed 5 acres in Great Briche from John son of Ralph Bygod and Joan in 1357,FF.

CRAY'S HILL, Ramsden Crays
Sasselin held *Winthelle* in 1086,DB: ?windy hill (above River Crouch). *(See Ramsden Crays).*

CREFFIELD ROAD, Colchester
Ralph *Creffeild* (died 1723), alderman, lived at Holly Trees. His widow's husband Charles Gray, MP, owned Colchester Castle. *(Fig. 85).*

CREPPING HALL, Wakes Colne
Crepinga was held by **Modwin** in 1086,DB: ingas, Creppa's peoples' settlement.

CRITTALL CLOSE, Silver End, Rivenhall
Francis *Crittall* (1860-1935) ran a window frame factory. He and his son Valentine (Baron Braintree) built social housing, a park and gardens catering for the workers' needs.

CROCKLEFORD BRIDGE, Ardleigh
In 1206,FF, Richard de Bovill acknowledged the right of Henry prior of St. Botulph of Colchester to all the land and wood of *Crockeresford*: croccere, potter's ford.

CROMPTON STREET, Chelmsford
Col. Rookes *Crompton* (1845-1940), founded the firm Crompton's in 1878, which pioneered electric generators supplying light to Buckingham Palace and designed 1st World War tanks. Closed 1969.

CROMWELL LANE, Maldon
Cromwell, 16th cent.,Bodl: crumb wielle, crooked stream/spring.

CROMWELL PARK, Felsted
Richard, son of Oliver *Cromwell* (1599-1658), Lord Protector during the Commonwealth, went to Felsted School. His father had married Elizabeth Bourchier of Felsted. *(Fig. 36).*

CRONDON, Stock
Cromden, 1213,ClR: crumb denu, winding valley.

CROPPENBURG WALK, Canvey
Joas *Croppenburg*, a Dutchman, helped finance the 'draining' and embanking of Canvey in the 17th century.

CROUCH STREET, Colchester
named from the Crossed or Crutched Friary founded *c.*1244 (site on the Maldon Road).

CROWSTONE AVENUE, Leigh-on-Sea
Tenement called Crowes, 1536. The Crow Stone marked the limit of jurisdiction of the City of London for payment of trade duties

CULVER STREET, Colchester
Colveraslane, 1335,Ct: culfre, a dove (?market).

CULVERT'S FARM, Boreham
Richeham was held by **Robert** Gernon whose tenant was Asso in 1086,DB: ?ric ham(m), narrow land-strip settlement or corner. Matilda quitclaimed to Richard de *Coleworde*, Northants, 12 acres in Boreham for 20 marks of silver, in 1253,FF.

CUNNINGTON ROAD, Braintree
John *Cunnington* wrote a history of Braintree in 1833.

CURLING TYE, Woodham Walter
Ralph Baynard held *Curlai* in 1086,DB: cweorn leah teag, mill clearing enclosure.

CURRY, Bradwell-on-Sea
William de la Mare quitclaimed the manor of *la Curie* for 14 marks yearly of Robert de Typpetot, 1262,FF: Old French curie, kitchin.

CUT-A-THWART LANE, Beeleigh
Uncertain but possibly Middle English thwert, cross at right angles, hence Cut-across.

CUTON HALL, Springfield
Geoffrey de Mandeville held *Keuentuna* in 1086,DB: ?cyf tun, settlement in a hollow (by River Chelmer).

CYMBELINE WAY, Colchester
Cunobelinos, king of Trinovantes tribe in Essex, died *c.*40 AD. His centre was probably at Gosbecks near Stanway. His sons, Togodumnus and Caratacus, opposed Emperor Claudius' invasion in 43 AD.

DACRE CRESCENT, Aveley
Thomas Lennard, 17th Lord *Dacre*, (1717-86), historian, remodelled Belhus (House).

DANE ROAD, Ilford
Frederick *Dane*, MBE, 1872-1955, first mayor of Ilford Borough Council in 1926.

DE BOHUN COURT, Saffron Walden
Humphrey de **Bohun**, earl of Essex, lord of Brecon, died in Saffron Walden in 1298. His son, also Humphrey, held Writtle, Pleshey and other estates in Essex, died 1322.

DE VERE ROAD, Earls Colne
(See Priory Street, Earls Colne).

DE VERE WAY, Dovercourt
Aubrey de Vere (from Ver, La Manche or Calvados) was Domesday holder of Dovercourt, 1086. *(Fig. 33).*

DEANE'S CLOSE, Harwich
Sir Anthony *Deane*, shipbuilder for the Navy in the 18th century, twice mayor of Harwich.

DEBDEN (ROAD), Loughton
Earl **Harold** gave *Tippedene* to found Waltham Holy Cross, 1062 (K813): Tippa's denu, valley. Tippaburne was the stream which rises on Jack's Hill. *Dupedene*, 1271,Loughton: deop denu, deep valley. *(Fig. 14).*

DEENE AVENUE, Hornchurch
Al *Deene* flew Spitfires from Hornchurch Airfield in 2nd World War.

DEFOE CRESCENT, Colchester
Daniel *Defoe* (1661-1731), novelist, set part of *Moll Flanders* in Colchester where he had a small estate Severalls.

DELVIN END, Sible Hedingham
Delvyn, 1576,Ct: dylf, a hollow, pit. *See Hedingham.*

DENGEWELL, Great Oakley
Daniswell, 1248,Ass: ?Denisc welle, Danish well. (A headwater of the Ramsey Brook).

DENNY AVENUE, Waltham Abbey
Sir Anthony *Denny* (1500-1549), Gentleman of the Privy Chamber to Henry VIII, leased Waltham Grange in 1540. His son Baron Edward of Waltham, 1604, was created Earl of Norwich in 1626.

DERES BRIDGE, Great Leighs
In 1200,FF, William Widecoc granted to Alured Gulle 10 acres in Welveld and 10 acres against *Deresbrege* for 4 shillings yearly: deor, bridge for animals.

DERHAM GARDENS, Upminster
William *Derham* (1657-1735), vicar of Upminster, FRS, published 'Astro-Theology' and 'Physico-Theology' in 1716, detailing 'Natural Variation', presaging Charles Darwin's 'Natural Selection' by 143 years. *(Fig.86, Fig. 87).*

86. The Rev. Dr. William Derham.

DE VERES ROAD, Halstead
Aubrey de Vere, 1086,DB, from Ver, La Manche or Calvados, was ancestor of the Earls of Oxford, landholders in Essex. *(Fig. 6, Fig. 33).*

DEVEREUX PLACE, Colchester
Walter *Devereux*, Earl of Essex (1541-1576), held several manors including Tolleshunt, Tollesbury and Goldhanger. First husband of Lettice Knollys *(see Dudley Court).*

DICKLEY HALL, Mistley
Robert Gernon held *Dicheleia*, 1086,DB: Dica's or dica leah, clearing between the streams.

DILBRIDGE ROAD, Colchester
Dilbridge, 1536,LP: possibly Old English dile, bridge near where vetch grew.

87. Title page of Physico-Theology, 1716.

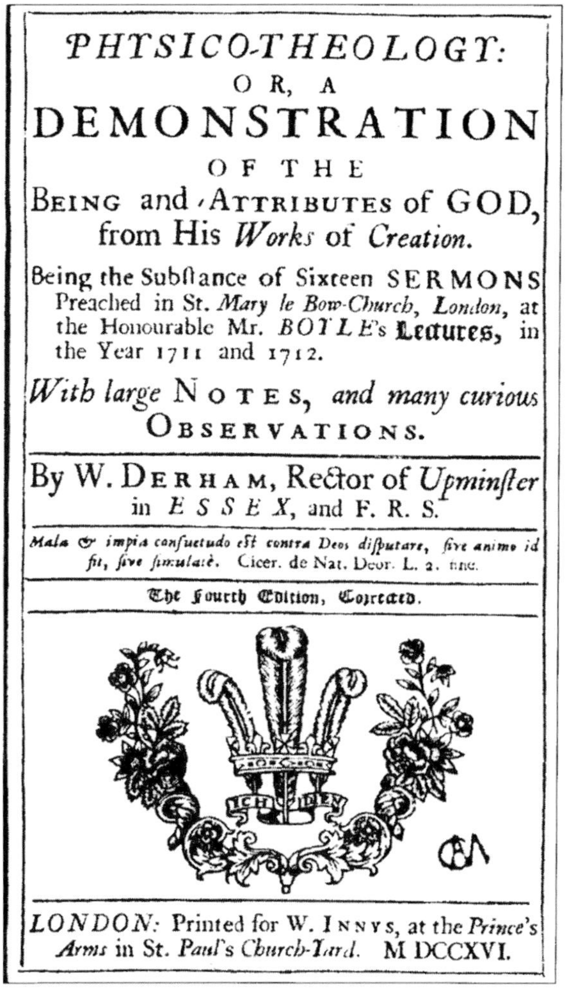

PHYSICO-THEOLOGY:
OR, A
DEMONSTRATION
OF THE
BEING and ATTRIBUTES of GOD,
from His *Works* of *Creation.*

Being the Subſtance of Sixteen SERMONS Preached in St. *Mary le Bow-Church, London*, at the Honourable Mr. *BOYLE*'s Lectures, in the Year 1711 and 1712.

With large NOTES, *and many curious* OBSERVATIONS.

By W. DERHAM, Rector of *Upminſter* in *ESSEX*, and F. R. S.

Mala & impia conſuetudo eſt contra Deos diſputare, ſive animo id fit, ſive ſimulatè. Cicer. de Nat. Deor. L. 2. fine.

The Fourth Edition, Corrected.

LONDON: Printed for W. INNYS, at the *Prince's Arms* in St. *Paul's Church-Yard.* M DCCXVI.

DISNEY CLOSE, Ingatestone
Rev. Dr. John *Disney* of Hyde House, 1746-1816, of the family from Norton Disney, Lincs, originally from Isingy near Bayeux. His son left considerable Roman and Greek collections to the Fitzwilliam Museum, Cambridge.
DOCWRA ROAD, Kelvedon
Miss Lucy and Miss Mary *Docwra*, social worker and temperance advocate, died 1914. Their family had lived here at Grays (House) for 160 years.
DOREWARDS AVENUE, Bocking
John *Doreward* of Doreward Hall, Speaker of House of Commons in 1399 and 1413.
DOROTHY SAYERS DRIVE, Witham
Dorothy Leigh *Sayers* (1893-1957), writer and author of *The Man Born to be King*, lived at 22 Newland Street.
DOUBLEDAY DRIVE, Heybridge
Herbert *Doubleday* (1867-1941) initiated the Victoria County Histories, of which the Essex series is now edited from Essex University.
DOWCHES GARDENS, Kelvedon
Was *Doucescroft*, a field with a dwelling, in 1417,MnAt.
DOWN HALL, Bradwell-on-Sea
Eudo the Steward held *Dona* and **Ranulf** Peverel held *Duna* in 1086: dun, hill.
DRIBERG WAY, Braintree
Tom *Driberg*, Baron Bradwell (1905-76), *alias* William Hickey, MP for Maldon and Barking, wrote for the *New Statesman*.
DRIFFIELD CLOSE, Feering
Walter Wren *Driffield*, curate of Feering for 50 years (died 1828).
DU CANE PLACE, Witham
Peter *Du Cane*, 18th century worsted merchant, extensive landholder including in Braxted, Tollesbury and Coggeshall.
DUCKETTS MEAD, Roydon
Ralph *Duckett*, gent of Roydon, died 1587, held land in Eltham, Kent, and tenements in Knightsbridge, London.
DUDDENHOE END, Elmdon
Duddenho, 1189-99,Dugd; in 1205,FF, Robert Leholt granted ½ virgate in *Dudeho* to Simon abbot of Tiletey for 4 shillings yearly: hoh, the hill-ridge of Dudda (above 400 foot, 120m contour).
DUDLEY COURT, Wanstead
Robert *Dudley* Earl of Leicester (1533-88), bought Wanstead, 'favourite' of Elizabeth I, married Lettice Knollys, countess of Essex *(see Devereux Place)*.
DUKE STREET, Chelmsford
Richard *Duk* through marriage acquired properties in Springfield and Brocholestrate, now Duke Street, c.1360 *(see also Threadneedle Street)*.
DUNKIRK ROAD, Burnham
From here, many small ships rescued troops from France in 1940.
DUNSPRING LANE, Barkingside
Led to *Downspring* (House), 1777,C&A.
DYNES HALL ROAD, Great Maplestead
In 1315 William *Deyn* and his son John of Halstede were granted by Robert Pechaud and William de Stubleye 2 messuages, land and wood in Halstede, Great and Little Mapeltrestede. His son Sir John *Dyne* of Bois Hall, Halstead, was MP for Essex 1310-38. The Hall built in 1580 was largely rebuilt in 1690 by Sir Mark Guyon, sheriff of Essex. *(Fig. 88)*.

EAGLE WAY, Great Warley
In 1812, the eagle standard was captured from 62nd French by 44th Foot Regiment posted from Warley Barracks.

88. Dynes Hall, Great Maplestead, c. 1926.

EALING BRIDGE, Harlow
*c.*1045, **Thurstan**, son of **Lustwine**, willed *Gildenebrigge* to his wife Æthelgyth (ECE 59), probably the ½ hide in *Herleua* of Count **Eustace**, 1086,DB: gildena brycg, bridge of the guildsmen (over Pincey Brook). *See Harlow.*

EASTBURY AVENUE, Barking
Eastbury House was the Tudor manor house built by Clement Sysley *c.*1560. burh, fortified/ manor.

EASTERFORD ROAD, Kelvedon
Estreford, 1213,Fees: easterra ford, eastern ford (through River Blackwater).

EAST HALL, Bradwell-on-Sea
Effecestra was held by **St. Valéry** Abbey in 1086,DB: castra, fort. Walle, 1212: refers to the fort. *(See Bradwell). (Fig. 6; Fig. 7).*

EASTLANDS, Bradwell-on-Sea
Eudo the Steward held *Landuna* in 1086,DB: land duna, lower land. *Estlond*, 1383,1426,Ipm.

EAST SWIN, Blackwater estuary
Atteswenne, 1310,Ct: (at the) swin, channel.

EDITH CAVELL WAY, Steeple Bumpstead
Trained as a nurse at the London Hospital. Matron of Red Cross Hospital, Brussels. Shot by the Germans in 1915 for assisting allied soldiers.

EDNEY COMMON, Writtle
Edenenhay, 1254,Takeley: gehæg, Eada's enclosure.

EDWARD BRIGHT COURT, Maldon
Edward *Bright* (1721-1750) lived at Church House, High Street. As a boy he delivered the post, then became a shopkeeper. Later in life he is reputed to have been the heaviest man in England at 42 stones. *(Fig. 89).*

EDWINS HALL LANE, Woodham Ferrers
Edwin Sandys born 1516, Bishop of London 1570-77, Archbishop of York 1577-1588, supporter of Lady Jane Grey, is said to have lived at the hall, formerly called Edwards Hall.

EGBERT GARDENS, Wickford
Saxon king of Wessex and first king of England (died 839).

ELEANOR CROSS ROAD, Waltham Abbey
Edward I set up crosses at the overnight resting stops bringing his wife Eleanor of Castile's body from Hadby, Nottinghamshire to Westminster Abbey in 1290.

89. Edward Bright of Maldon, c. 1750

ELDER STREET, Wimbish
Hylderestret, 1393,Ct: ellern straet, street by the elder-trees.

ELDRED AVENUE, Colchester
John *Eldred* (1552-1632), merchant, director of East India Co., lived at Olivers, Stanway.

EMANUEL WOOD, Little Chesterford
Count **Alan** held *Monehala* in 1086,DB. A[lan the Black I, earl of Richmond] granted *Monehale* to Bury St. Edmunds monastery *c.*1090: healh, Mana's corner. *See Chesterford.*

ENGLANDS, Little Warley
*c.*1070, **William I** granted *Gyddesdune* to 'Deorman', ? a Colchester moneyer, (ECE 72; EAH 16, p44). Westminster Abbey held *Geddesduna* in 1086,DB: dun, hill.

ENGLAND'S LANE, Loughton
Associated with George *England*, 1633,Harl. Formerly Ree Lane (1404,Loughton) from *æt thære ea*, at the stream (which flows southeast under Pyrles Lane, *qv*, to the River Roding).

ENDLEBURY ROAD, Chingford
Walter *Entirby* received lands in Chingford from Alice, widow of John Bowyer, in 1435,TEAS2iii,183.

ESDAILE GARDENS, Romford
Sir James *Esdaile* lived at Harwood Hall, Upminster, 1782.

ETHELBURGA ROAD, Harold Wood
First Abbess of Barking monastery, sister of St. **Erkenwald** *(qv)*. Died 676.

ETHELBERT ROAD, Ashingdon
King **Æthelbert,** died 616, of Kent sponsored Sabert king of the East Saxons to build St. Paul's, London. A king of the East Angles of the same name was murdered by Offa of Mercia in 794, and buried at Hereford.

EUDO ROAD, Colchester
Eudo Dapifer, steward to **William I** and II. Died in Préaux, Normandy, in 1120.

FABIANS CLOSE, Coggeshall
Robert *Fabion* (died 1513) of Theydon Garnon, Alderman of the City of London. A branch of the family lived in Coggeshall.

FACTORY LANE, Halstead
Courtauld built workers' houses for his mill factory, 1872. *(See Courtauld Road, Braintree).*

FAIRFAX ROAD, Colchester
Sir Thomas *Fairfax* led the Parliamentarian Army against the Royalists defending Colchester, 1648. *(Fig. 90).*

FAIRLOP, Ilford
Fair Lop Oak, 1738: ancient 36 foot diameter oak lopped in 1767 by George III's gardener for a coffin for Daniel Day of Wapping who had held yearly parties for his employees beneath it. It was cut down soon after Hainault Forest was freed from forest laws in 1851. *Essex Naturalist*,viii,94. *(Fig. 18. Fig. 91).*

FANSHAWE CRESCENT, Barking and Dagenham
Henry *Fanshawe* (1506-68) purchased the Manor of Jenkins in Barking in 1557. Parsloes was bought by Sir William *Fanshawe*, Auditor to the Duchy of Lancaster, in 1619.

FANTON HALL, North Benfleet
In 969,(CS 1264), King **Edgar** confirmed the grant of *Fentune* to Westminster Abbey. In 1065, (K824), King **Edward** confirmed the grant of *Parva Fentune* to Westminster Abbey who held *Phantuna* in 1086,DB.: fenn tun, fen settlement. *See Benfleet.*

90. 'Cromwell pypeth unto Fairfax' (See Fairfax Road).

FELL CHRISTY, Chelmsford
Fell Christy went into partnership with his father James, of a family of millwrights, to open an engineering works in 1858, in 1897 Christy and Norris making agricultural machinery.

FILLIOLL CLOSE, Chelmsford
Johannes *Filiol* represented Essex as MP in 1290.

FINGRITH HALL, Blackmore
William I held *Phingheria* in 1086: inga rith, Fin's peoples' stream. Headwater of Ewson's Brook.

FINUCANE GARDENS, Hornchurch
Brendan 'Paddy' *Finucane*, DSO, DFC, flew Spitfires from Hornchurch airfield in the 2nd World War.

FISHERS GREEN, Waltham Abbey
c.700, (Kelly 65a), *Stanhemstede* was southern boundary of King Suebred's grant of land at

91. The Fairlop Oak at Hainault in 1817.

Nazeing: settlement place by a stone. Associated with John *Fisshere*, 13th century,Wm. *See Waltham.*

FITZWALTER PLACE, Chelmsford

Posthumously in 1667, Sir Henry Mildmay was recognised as 15th Baron *Fitzwalter* through his grandfather the 11th baron. *(Fig. 92).*

FITZWALTER PLACE, Great Dunmow

Robert **Fitzwalter** (?1170-1235), Lord of Dunmow and Baynard's Castle in London, one of the barons confronting King John resulting in his signing Magna Carta in 1215.

FOREST GATE, Wanstead

Documented in 18th century. The gate to Epping (Wanstead) Forest in Woodgrange Road was taken down in 1883. *(Fig.93).*

FOSSETTS LANE, Fordham

Possibly from Middle English *fosse,* ditch: Excavation has found a Roman lead coffin.

FOSTER STREET, Harlow

Robert and Alice Serle of *Forsterstrete* in Harlow sold a messuage and 7 acres in Matching and Sheering, 1440,FF; William le *Forester*, 1327,SR.

92. Henry Mildmay, Lord Fitzwalter, memorial in Chelmsford cathedral.

93. *Wanstead Flats.*

FOUCHERS, East Horndon
Ginga was held by **Ranulf brother of Ilger** in 1086,DB: ge ingas, dwellers of the district. Henry fitz *Fulcher* quitclaimed land in Ginges to Warin fitz Warin for 10 marks of silver, 1203,FF. *See Horndon.*

FOULTON HALL, Ramsey
Fuletuna was held by **Swein** in 1086,DB: ful tun, dirty settlement.

FOUR WANTZ, Ongar and Upminster
Middle English 'wente', a path, road, hence cross-roads.

FRAMBURY LANE, Newport
See Gaces Acre, Newport

FRANKS, Great Warley
Warleia was held by **Swein** in 1086,DB. *Frank* de Scodlaund held 1 messuage, 2 caracutes by rent of 1lb of cumin, now 10 pounds sterling, 1262,FF. *See Warley.*

FRESHWELL STREET, Saffron Walden
Probably 'fresh spring', lying close to the course of the The Slade. *Fross(c)ewella*, 1086,DB: forsc wielle, frogs' spring, probably refers to another Freshwell on River Pant.

FRIARS LANE, Maldon
A Carmelite friary founded in 1292 stood here, the gatehouse behind 40 High Street.

FRIARS WALK / PLACE, Chelmsford
Site of a Dominican Priory in 13th century, until dissolved in 1538 by Henry VIII. *(Fig. 94).*

FRISTLING HALL, Buttsbury, Stock
Barking monastery held *Festingas* in 1086,DB: ingas, settlement of Fyrstel's people. *See* ***Erkenwald.***

FROWICK, St Osyth
Froruuica was held by Count **Eustace** in 1086,DB: ?Scandinavian vik, Frothi's sea-inlet (from Flag Creek).

FRYERNS, Basildon
ffreryn, 1469,MnAt. In 15th cent., held by brothers and Knights Hospitaller of St. John.

FULFEN WAY, Saffron Walden
Named from Fulfen Slade. Old English ful fenn, dirty marshland; slaed, valley.

FULLERS CLOSE, Waltham Abbey
Robert *Fuller*, last abbot of Waltham Abbey, dissolved in 1540. Thomas *Fuller* (1608-1661), vicar of Waltham Abbey, wrote several church histories, and was beneficiary of a fine library at Copt Hall.

94. A sculpted Dominican friar in Chelmsford.

GACES ACRE, Newport
John *Gace* in 1436 held land called *Franebyry* (Great, Little *Frambury*, Tithe 1842; *Frambury* Lane.)

GALE STREET, Dagenham
Galesstrete, 1452,AddCh; associated with William le *Gal*, 1285,Ass.

GALLEYHILL ROAD, Waltham Abbey
Le Galwehill, 1414,Ct: gealga, gallows.

GALLEYWOOD, Great Baddow
Gauelwode, 1250,CottCh: gafol wudu, taxed wood.

GAMUEL ROAD, Walthamstow
Thomas *Gamuel*, 1642, local benefactor.

GANTS HILL, Ilford
Associated with Ralph le *Gant*, who was Steward of Barking Abbey, 1321.

GASCOIGNE ROAD, Colchester
Sir Thomas *Gascoigne* defended Colchester for the king against Colonel Fairfax in 1648.

GAUDEN ROAD, Bocking
John *Gauden* (1605-62) was Dean of Bocking, chaplain to Sir Richard Rich (earl of Warwick), then Bishop of Exeter.

GAUNT'S END, Elsenham
Peter de Gandavo [*Ghent*] quitclaimed to Henry and Isabel de Ghent manor of Elsenham in 1268,FF.

GEPPS CLOSE, High Easter
Rev. Edward *Gepp* of High Easter (1855-1929), historian and schoolmaster at Felsted, wrote *An Essex Dialect Dictionary* in 1923.

GERNON CLOSE, Chelmsford
Robert *Gernon*, "the moustached", Domesday (1086) holder of extensive manors in Essex, the chief at Stansted. William of the family Monfiquet (from Normandy) took his estates.

GIDEA PARK, Romford
Gidiehall, 1258,WmC: gydig, mad-venture hall, or stream-name (for upper reaches of Ravensbourne).

GILBERD ROAD, Colchester
William *Gilberd* (1544-1603) of Tymperleys Colchester, president of the Royal College of Physicians, and the first to publish that the Earth was a magnet, in 1600, setting the science of electricity. Buried in Holy Trinity, Colchester.

GILBEY COTTAGES, Elsenham
Sir Walter *Gilbey* (1831-1914) founded the firm of wine merchants. He came to live at Elsenham and endowed a Lectureship in Agriculture at Cambridge.

GLAZENWOOD, Bradwell-by-Coggeshall
*c.*995, (K699), **Æthelric** willed *Glæsne* to his wife then to Christ Church Canterbury, but in 1086,DB, this outlier of Copford manor was held by **Robert** Gernon: glisian, bright spot or Celtic glesin, grassy clearing.

GLEBE ROAD, Hornchurch
Glebe: land belonging to the church and so usually exempted from payment of Tithe dues.

GODFREYS MEWS, Moulsham
Henry *Godfrey*, ropemaker in Moulsham, from 1828.

GOLD STREET, Saffron Walden
Charles *Gold* of The Limes Birchanger, Liberal MP for Walden 1895-1900, knighted 1906.

GOLDING WAY, St Osyth
Benjamin *Golding* (1793-1863), born at St. Osyth, physician. In 1818 by Haymarket Theatre, he founded The West London Infirmary, which in 1827 became Charing Cross Hospital.

GOLDINGHAM HALL, Bulmer
Goldingham was held by **Robert Malet** in 1086,DB: ingaham, settlement of Golda's people.

GOLDINGS HILL, Loughton
Golden Hill, 1777,C&A, at the junction with Clays Lane, *qv*. John *Goldyng* paid 3/1d tax in Loketon, 1319,Loughton.

GOODMAYES, Ilford
Goodmayes, 1456, ERxxxii: John *Godemay* paid tax at Ilford in 1319.

GOOSHAYS, Romford
Gosaye, 1334,Fine: gos gehæg, goose enclosure.

GOSBECKS, Stanway
Gosebekks, *c*.1540,LP: associated with Roger de *Gosebek* from Gosbecks near Needham Market, Suffolk, 1254,Ass.

GRANGE HILL, Coggeshall
The grange farm of the abbey. The barn has been tree ring-dated to 1240.

GRANSMORE GREEN, Felsted
Grendelesmere, 12th cent, Cartulary of Caen Abbey: The monster *Grendel* killed by the warrior Beowulf who fought a mortal fight at the mere "overcast with dark, crag-rooted trees that lie in frost-hoary groves", 8th-century poem.

GRAY ROAD, Colchester
Charles *Gray* (1696-1782), Fellow of the Royal Society, MP, barrister, lived at Hollytrees House and was given Colchester Castle as a wedding present.

GRIMSTON WAY, Harwich
Harbottle *Grimston*, bart (1603-85), MP for Harwich, High Sheriff. For his associations with Charles I he was imprisoned in The Tower. After the Restoration he was elected Speaker of the House of Commons.

GUBBINS LANE, Harold Wood
John *Gobyon* gifted the manor of Uphavering in 1467,AD.

GUBBION'S HALL, Great Leighs
Gobions, 1376,Cl: Thomas and Blanche *Gobioun* in 1317,FF, quitclaimed 1 messuage, 172 acres, 4½ acres of meadow, 30 acres of pasture, 8 acres wood from John de Giney of Little Bardfield for £100.

GUITHAVON ROAD, Witham
A concocted name of 18th century supposedly refers to the 'avon' or river of Witham.

GURNEY BENHAM CLOSE, Colchester
Sir William *Gurney Benham* (1859-1944) was Mayor of Colchester and High Steward of Essex. He and his brother Charles were authors of distinction.

GURNEY CLOSE, Barking
Samuel *Gurney*, a Quaker banker, was prominent with Elizabeth Fry, his sister-in-law, in penal reform. He was buried in Barking in 1856.

GUTTERS LANE, Broomfield
A detached part of Chelmsford until transferred to Broomfield in 1889.

HACTON BRIDGE, LANE, Upminster
Haketon, 1310,HPD: haca tun, settlement at the tongue of land. (River Ingrebourne).

HADRIAN'S CLOSE, Witham
Roman Emperor, AD 117-138. Visited Britain in AD 122.

HAINAULT FOREST
Henehout, 1221,Cl: higna holt, community wood (of the nuns of Barking monastery). *(Fig. 18)*.

HALES FARM, Purleigh
? Hugh de Montfort held *Halesduna* in 1086,DB: Hægel's dune, upland?. *(Fig. 6)*.

HANGING HILL LANE, Shenfield
Derived from Old English 'hangra', a wooded slope.

HARCAMLOW WAY, Harlow and Newport
Leads from HARlow to CAMbridge. It was opened in 1980 to mark the 10th anniversary of West Essex Ramblers' Association.

HARLINGS GROVE, Chelmsford
John *Harling* arrived in Chelmsford in *c*.1407 and with his son Guy built a speculative estate east of New Street. In 1424 a house 'Sadlers' was sold by John Mason to Guy Harling. The ancient house called 'Guy Harlings' was rebuilt *c*.1720.

HARMAN CRESCENT, Chingford
Edward & Henry *Harman*, London merchants, surrendered land to Sir Robert Ainsley in 1845,TEAS2iii,188.

HARRISON DRIVE, Braintree
Dr. John *Harrison* (1857-1929) gained notoriety by his practical jokes played on his patients.

HARSNETT ROAD, Colchester
Samuel *Harsnett* (1561-1631), son of William Halsnoth, baker of St. Botolphs. He was master of Colchester Grammar School, rector of Stisted, bishop of Norwich, then archbishop of York. He founded Chigwell School where he had been vicar.

HARVEY ROAD, Wivenhoe
Sir John Martin-*Harvey* (1863-1944), actor-manager, was born at Wivenhoe. He worked with Sir Henry Irving at The Lyceum, London.

HARVEY WAY, Hempstead
Sir William *Harvey* (1578-1657), physician to St. Bartholomew's Hospital, discoverer of the circulation of the blood, published *De Motu Cordis* in 1628. Buried in the family vault at Hempstead. *(Fig. 95).*

HARVEY WAY, Saffron Walden
Gabriel *Harvey* (*c.*1550-1630), born at Saffron Walden, poet, 'the father of the English hexameter'. His brother Richard (*c.*1560-1631) was an unsuccessful astrologer and astronomer.

HASSENBROOK, Stanford-le-Hope
Odo held *Hasingebroc* in 1086,DB: broc, stream of Hassa's people (which runs into Mucking Creek).

HAWKESBURY BUSH, Fobbing
Hauechesberga, 1166,P: Heafoc's or hawks' hill (south of Dry Street overlooking the Thames valley).

HAWKWOOD CRESCENT, Chingford
Middle English *halke*, a corner of land (extending into Waltham Holy Cross).

HAWKWOOD ROAD, Sible Hedingham
Sir John de *Hawkwood* (1340-94) fought in Edward III's Hundred Years War in France. Dying in Italy his memorial (?tomb) is in Sible Hedingham where he was born.

HEADLEY CHASE, Brentwood
Lord *Headley*, lord of the manor of Brentwood. The Headley Arms was frequented by soldiers from the nearby Warley Barracks.

HEENAN CLOSE, Barking
Cardinal John *Heenan*, 1905-75, born in Ilford, Roman Catholic Archbishop of Westminster.

HEMNALL STREET, Epping
Hemenhales, 1343,Cl.; in 1376,FF, Thomas Tirell, chivaler, and Alice held the manor of *Hemenales in Dounham* etc. of the chief lords. Henry de

EXERCITATIO
ANATOMICA DE
MOTV CORDIS ET SAN-
GVINIS IN ANIMALI-
BVS,
GVILIELMI HARVEI ANGLI,
Medici Regii, & Professoris Anatomiæ in Collegio Medicorum Londinensi.
FRANCOFVRTI,
Sumptibus GVILIELMI FITZERI.
ANNO M. DC. XXVIII.

95. Title page of De Motu Cordis, *1628.*

Emhal, 1248,Ass, may have come from Hempnall, Norfolk.

HENGIST GARDENS, Wickford
Hengist, legendary brother of Horsa, leaders of Saxon mercenaries who fought the British and settled in Kent *c.*449 AD.

HERONGATE, Ingrave
In 1232,FF, Ralph de Ginges granted Richard abbot of Strafford common and pasture for 240 sheep in Ginges from *Findegodehurne* towards the north for 2 marks: hyrne, corner + gate (to the forest). John Fyndegod, regarder of Forest of Essex, 1365.

HEWETT ROAD, Barking and Dagenham
In 1830s, Samuel *Hewett* operated a large fishing fleet from Barking, introducing refrigeration and fast cutters to bring the catch home before his rivals.

HEYBRIDGE, Ingatestone
Pont de Heybrigge, 1323,For: gehæg, brycg, bridge by the enclosure/farm (tributary of River Wid). *(Fig. 96).*

96. *The 18th-century farmhouse, Heybridge, Ingatestone.*

HICKS WAY, Sturmer
William *Hicks* (1788-1874), rector, served on the *Conqueror* at the Battle of Trafalgar.

HIGHAMS PARK, Walthamstow
Peter of Valogne held *Hecham* in 1086,DB: heah ham, high settlement (above River Lea).

HOFFMANNS WAY, Chelmsford
Ernest Gustav Hoffman's factory, opened 1898 in New Street, manufactured roller bearings until closure in 1968.

HOLMAN ROAD, Halstead
William *Holman*, pastor of the Independent Meeting House and Essex antiquary, died 1730.

HOLCOME ROAD, Ilford
Holcome Ingleby, 1854-1926, MP for King's Lynn, Norfolk, gave land at Valentines. *(Fig. 105).*

HOLMES ROAD, Halstead
Elizabeth *Holmes* (1706-83), a member of a prosperous family, has an inscription in the church.

HOLYBREAD LANE, Little Baddow
Produce from the field was endowed for purchase of Holy Communion bread.

HONEY LANE, Waltham Abbey
Honylane, 1408,Ct: perhaps 'sticky soil', 'productive land', or where bees were kept.

HONYWOOD ROAD, Colchester
Sir Robert *Honywood* of Markshall, Coggeshall (1587-1666) led Cromwell's Parliamentary Army and was involved in the Siege of Colchester in 1648.

HOOD LODGE, Wanstead
Thomas *Hood* (1799-1845) of Lake House, Wanstead, poet and contributor to the *London Magazine.*

HOPKINS MEAD, Chelmer Village
Matthew *Hopkins* (hanged in 1647) lawyer of Manningtree was self-proclaimed Witchfinder General. In trials all around London he zealously prosecuted more than 60 women who were hanged for witchcraft. *(Fig. 97).*

HORSEHAM HALL, Helion Bumpstead
Ersham was held by St. Martin's Abbey, Battle (Sussex) *(Fig. 44)* in 1086,DB: herse ham, ?hilltop settlement. The present hall is at 106m OD.

HOUBLONS HILL, Coopersale, Epping
Jacob *Houblon* married Susannah Archer of Coopersale *c.*1769 and moved to Hallingbury

97. Matthew Hopkins, Witch Finder Generall.

Place, Great Hallingbury. He served with the Hertfordshire Militia, dying in 1783. John Archer *Houblon* (died 1891) rebuilt the church, enclosed the Forest and improved the garden and village *(Fig. 42)*.

HOUSHAM HALL, Matching
Ouesham was held by William of **Warenne** in 1086,DB: ham, Ufe's settlement. *See Matching.*

HOWE HALL, Finchingfield
Richard son of Count Gilbert held *Weninchou*, 1086,DB: hoh, promontory-hill (at 85m OD).

HOWARD ROAD, Ilford
Luke *Howard* (1772-1864), FRS, of Ilford, meteorologist, climatologist and chemist.

HOWARD ROAD, Saffron Walden
Lord Thomas *Howard* (1561-1626), created Lord Treasurer in 1614, completed Audley End but after imprisonment was ruined.

HOWBRIDGE HALL ROAD, Witham
Bishop of London held *Hobruge* in 1086,DB: hoh brycg, bridge on a spur of land.

HOWE STREET, Great Waltham
Lohou was held by Roger Marshal in 1086,DB: loh hoh, ?place on a promontory. (It stands on a curve of River Chelmer). *See Waltham, Great.*

HOWEGREEN, Purleigh
Christ Church Canterbury held *le Hoo*, 1310,TEASxx,174: hoh, hill-spur (above tributaries of Lime Brook).

HUBERT ROAD, Colchester
Hubert, burgess of Colchester, MP for Colchester Borough in 1295.

HULLBRIDGE, Hockley
Whoulnebregg, 1375,Ct: bridge over river Huolne (former name of Crouch, *qv*). Middle English wholve, water-passage.

HUNT ROAD, Earls Colne
Reuben *Hunt* (1836-1927), engineer, enlarged the grammar school and started the Atlas Engineering Works.

HUNTER AVENUE, Brentwood
William *Hunter*, burned at the stake (in Ingrave Road) for his non-conforming faith in 1555.

HUNT'S HALL, Pebmarsh
William of **Warenne** held *Polheia* in 1086: pol (ge)haeg, enclosure by a pool. Simon le *Hunte* held one messuage, one mill, and land in *Pebeners* in 1312,FF.

HYDE FARM, Tolleshunt D'Arcy
One hide in *Tolesberia* was held by **Ranulf** Peverel from Barking monastery in 1086,DB. *See Tollesbury.*

HYTHE HILL, Colchester
Hethstret, *c*.1330,Ct: Old English hyth, landing-place (on River Colne).

IBBETSON PATH, Loughton
Sir Henry Selwin-*Ibbetson*, Lord Rookwood (1826-1902), MP for West Essex 1868-84 and Epping 1884-92, Deputy Lieutenant, piloted a bill for opening Epping Forest to the public, 1878.

ICENI WAY, Colchester
The Iceni tribe in Norfolk and north Suffolk led by Queen Boudicca which laid waste to Colchester and London in 60/61 AD in the absence of the Roman legions in Wales.

ICKNIELD WAY, Great Chesterford
In 903, (CS 601), King **Edward** granted land at Hardwell in Compton Beauchamp (Berkshire) abutting *Icenhilde weg* to his fasallus Tata; in Essex in 1207,FF, Constancius de Heidene (Heydon) and Aleis granted Roger le Cornur 1 acre abutting *Ykenild* road. This trading route has been in existence probably for 5000 years since the Neolithic Age, connecting Norfolk with the West Country. The name may be a survivor of the language spoken before British-Celtic from around ? 2500 years ago, or OE hylte, Ica's wood.

ILTNEY FARM, Mundon
Count **Eustace** held *Eltenai* in 1086,DB: Elta's eg, low-lying land.

INGELRICA AVENUE, Hatfield Peverel
Ingelrica (died 1100) was mother to a son of **William I**. She married **Ranulf** Peverel, Domesday landholder. She founded a college for secular canons at Hatfield.

INGLEBY ROAD, Chadwell
Ingleby, one of the 4 manors created in the 12th century by the Bishop of London.

INGLEBY ROAD, Ilford
Sarah *Ingleby*, 1823-1906, President of Ilford Philanthropic Society, lived at Valentines. *(Fig. 105).*

INSKIP ROAD, Barking & Dagenham
James *Inskip* (1868-1949), vicar of Leyton, later bishop of Barking in 1919.

JACKSON ROAD, Clacton
William Parry *Jackson*, 1805-1881, of Woolwich, director of Steam Packet Co., opened the pier and The Royal Hotel in 1872.

JACQUES HALL, Bradfield
Roger de Raismes *(Fig. 6)* held *Manestuna* in 1086,DB: tun, Mann's settlement. Indenture between *Jakys* of Reynsfourgh of Bradfyld and William Dale squire of Dale Hall, Lalford, with Henry Chalys of Lalford, tanner, to purchase a tenement Perkynnys in Lalford, 1507,AD.

JAYWICK, Clacton
Clakyngewyk, 1438,MnAt: jay bird-sound + wic, farm or (Scandinavian) vik, bay, inlet.

JEKYLLS, Finchingfield
Phincingefelda was held by Count **Alan** in 1086,DB. Robert Jukel, 1235,FF. *See Finchingfield.*

JERICHO PRIORY, Blackmore
Jerico, 16th century. A familiar name for a distant place where Henry VIII could go "when he had a mind to be lost to his courtesans" far from London to visit Elizabeth Blount, mother of his son Henry Fitzroy.

JOHN TIBAULD COURT, Steeple Bumpstead
John *Tibauld* burned at the stake in 1527 for non-conformity and reading the Bible at Bower Hall.

JOSSELIN CLOSE, Earls Colne
Ralph *Josselin* (1618-1683), Puritan vicar of Earls Colne and diarist.

JOHNSON ROAD, Great Baddow
John *Johnson*, 1732-1814, built Chelmsford Shire Hall and Stone Bridge over River Can.

JOHNSTON ROAD, Woodford
Andrew *Johnston* (1835-1922), MP, Verderer of Epping Forest, first chairman of Essex County Council.

KATHERINE ROAD, East Ham
Katherine *Fry*, eldest daughter of Elizabeth Fry, prison reformer, who lived at Plashet House, demolished *c.*1883 *(see Plashet Road).*

KEEBLE PARK, Maldon
In 17th and 18th centuries, James *Keeble* and the family ran a large fleet of ships which transported grain to London.

KEERES GREEN, Aythorpe Roding
Richard and Margery *Keere* claimed a messuage, 29 acres of meadow in Abbots Roding of Hugh and Alice de Kyngeston in 1322,FF.

KEIR HARDIE WAY, Barking
James *Keir Hardie*, first Labour MP, for West Ham South 1892-5.

KEMPE ROAD, Finchingfield
William *Kempe* of Spains Hall (1555-1628) performed penance by remaining silent for 7 years. He was patron to Stephen Marshall, non-confirming vicar of Finchingfield.

KENNINGTONS, Aveley
William de **Warenne** held *Kelituna* in 1086,DB: Cylla's ingtun, settlement place.

KINCAID ROAD, St Osyth
General *Kincaid*-Smith lived at The Priory *c.*1930.

KINGS HILL, Loughton
George *King*, farrier in 17th century, worked here.

KITCHEN HALL, Harlow
Walla was held by **Richard** son of Count Gilbert in 1086,DB: wall, wall. *Kechenhall,* 1327,SR: supplied the Bury St. Edmunds abbot's kitchens. *See Harlow.*

KNIGHTON LANE, Woodford
*Knighton Wood, c.*1640,ERxiv: of Knights of St. John of Jerusalem.

LAKESIDE, Thurrock
A huge shopping mall built in a quarry in the 1980s alongside water-filled gravel pits.

LARKIN CLOSE, Brentwood
John *Larkin,* 1850-1926, Brentwood author, wrote *Fireside Talks about Brentwood,* and other works.

LAMMAS WAY, Wivenhoe
hlaf-mæsse=loaf mass, Lammas: annual festival (1st August) when first fruits of harvest were offered. One or more fields may have been kept for this purpose, e.g. Church Bread field, Felsted.

LASHLEY, Lindsell
Richard son of Count Gilbert held *Lacelea* in 1086,DB: lache leah, wood/clearing by a slow stream (a tributary of Stebbing Brook).

LATTON HALL, Latton
Lattuna was held by **Peter** of Valognes in 1086,DB. *See Latton.*

LAUND'S, Ashen
William de la *Lande,* Dugd vi,*c.*1090; *La Lande,* 1322,BL: Old French launde, glade in woodland.

LAUNDERS, Rainham
Before 1066 a priest held ½ hide in *Raineham,* then Hagebern in 1086,DB. *See Rainham.*

LAUNDERS LANE, Wennington
les Laundres, 1365,Ct (in Rainham): associated with John *Launder,* 1503,Ipm.

LAWFORD LANE, Writtle
Lollefordstrat, 1376,Grieve, was the king's highway from Writtle to Chelmsford in 1292. The ford crossed Roxwell Brook. *See Bishops Hall Road.*

LAWLING IN LATCHINGDON
In 991, ealdorman **Brihtnoth** granted *Lallinges* to Christ Church Canterbury (ECE 26). His widow Ælfflaed willed *Lellinge* to ealdorman **Æthelmær** *c.*1000,(CS 1289). King **Æthelred** confirmed 10 hides at *Lellincge* to found Eynsham Abbey, Oxfordshire, in 1005 (K714). Canterbury, **Eudo** the Steward and **Ranulf** Peverel held *Lalinge* in 1086,DB: stream name (Celtic)? *(See also Latchingdon).*

LEATHART CLOSE, Hornchurch
James *Leathart,* member of 54 Squadron, flew from Hornchurch Airfield in 2nd World War.

LEGG STREET, Chelmsford
Richard *Legg, c.*1620, owned a croft here.

LENNARD ROW, Aveley
Thomas *Lennard,* Lord Dacre (1717-86) restored and updated his house Belhus in Aveley.

LEVERTON WAY, Waltham Abbey
Thomas *Leverton,* architect, founded by bequest the school here in 1824.

LINDSEY STREET, Epping
Lincedeheie, 1200,P: hlinc heafod (ge)haeg, enclosure on the hill-top.

LISLE ROAD, Colchester
In 1648, Sir George *Lisle* defended Colchester for the Royalists against the Parliamentarians. On its surrender he was executed by shooting.

LISTER STREET, West Ham
Dr. Joseph *Lister* (1837-1912), Order of Merit, was born at Upton House, West Ham. He introduced aseptic practices to surgery thus saving many lives from infection.

LITTLE THORPE, Southchurch
Odo held *Torpeia* from **Swein**, 1086,DB: Old English or Old Scandinavian throp, thorp, settlement.

LITTLEY GREEN, Great Waltham
The Prior's wood of *Littlehey,* 1247,FF: (ge)haeg, little enclosure. *See Waltham, Great.*

LIVINGSTONE CLOSE, Ongar
Dr. David *Livingstone* (1813-1873) trained as a missionary at Ongar before making expeditions to seek the sources of the Nile.

LONGHOUSE ROAD, Chadwell
Longhouse, an early medieval house divided for livestock at one end and for dwelling at the other. Longhouse was one of 4 manors created in 12th century by the Bishop of London.

LOOMPITS WAY, Saffron Walden
lam pytt, loam diggings pit.

LOVERS LANE, Great Bentley and Harkstead
Suitably distant from other distractions.

LOXFORD LANE, Ilford
Loxfordstrete, 1456,ERxxxii; ford through Loxa stream (now Ilford Lane over Loxford Water) + street.

LUCAS ROAD, Colchester
Sir Charles *Lucas* (1613-48), son of Sir Thomas of St. John's, Colchester, led the Royalist defence of Colchester and was executed by shooting after the Siege.

LUFF WAY, Frinton and Walton
Ernest *Luff*, evangelist, lived at Frinton and opened a home for the elderly in the 1930s.

LUTHERS CLOSE, Kelvedon Hatch
Mr *Luther* of Kelvedon Hatch was a strong supporter of the Parliamentarians in the Civil War. His namesake John *Luther*, MP, 1786, lived at Myles's.

LUXBOROUGH, Chigwell
William de *Lughteburgh* (Loughborough, Leics) quitclaimed 1 messuage, 122 acres etc in Chigewell to Eva Werplok for 100 marks of silver, in 1316,FF.

LYNGS LANE, Loughton
Lynges Lane, 1460,Loughton; Robert *Lyng* was a defaulter tenant in Luckton, 1538,Loughton.

MAIDENBURGH STREET, Colchester
Maydeneburghstraat, 1372,Ct: maidens' fort. There is no certain evidence for a fort here though it was the site of the Roman theatre. Perhaps a derogatory quarter of the town.

MALAN SQUARE, Hornchurch
Adolph *Malan*, Spitfire pilot in 2nd World War, flew from Hornchurch Airfield. *See Squadrons Approach.*

MALGRAVES, GREAT, Horndon on the Hill
*c.*1277 John *Malegreff* witnessed a charter concerning Dunton Malgraves, TEASxvii,101.

MANDEVILLE WAY, Basildon, Broomfield & Kirby Cross
Geoffrey de *Mandeville*, (perhaps Mandeville, Eure, France), Domesday (1086) holder of extensive lands in Essex. His grandson, Earl of Essex, 1140, had castles at Pleshey and Saffron Walden, mortally wounded, 1144.

MANGAPP CHASE, Burnham
Uncertain but possibly O.E. (ge)maene gappe, opening in the fence to common land.

98. Guglielmo Marconi.

MANSFIELD HILL, Walthamstow
Manfeld: O.E. (ge)maene, land held by common possession.

MARCONI ROAD, Chelmsford
Guglielmo *Marconi* (1874-1937) pioneered electrical impulses sending the first trans-Atlantic wireless Morse signal in 1901. *(Fig. 98)*

MARDEN ASH, High Ongar
*c.*1045, **Thurstan**, son of **Lustwine**, willed a half hide near *Meredene* to Merewine (ECE 59). **Ralph** Baynard held *Angra* from Count **Eustace** in 1086,DB: (ge)mære denu, boundary valley (between High and Chipping Ongar). *See also Ongar and Ashlyns.*

MARDYKE ROAD, Harlow
(ge)maere dic, boundary ditch and bank.

MARK HALL, Latton
Count **Eustace** held *Lattuna*. Adelulf de *Merc* (Pas-de-Calais) was his undertenant, 1086,DB. *See Latton.*

MARK ROAD, Tillingham
Old English mearc, boundary, on north of the parish.

MARKS, Great Dunmow
Adelulf de *Merc* (from Marck, Pas-de-Calais) held from Count **Eustace** *Dommauua* in 1086,DB. *See Dunmow.*

MARKS GARDENs, Braintree
Merkes, 1430,Ct; associated with William de Mark, 1254,Morant.

MARKS GATE, Dagenham
Old English mearc, boundary, on edge of the forest.

MARKS HALL, Margaret Roding
Hamo the Steward held *Rodinges* in 1086,DB. Gilbert de *Merc* claimed from Thomas de Newell one virgate in Roinges for 20 shillings in 1201,FF. *See Roding.*

MARLPITS ROAD, Woodham Walter
Marl, a mixture of clay and lime, has been used to fertilise the soil since the Iron Age.

MARSHALLS, Braintree
John de Borham to hold 1 messuage etc. from Robert son of Richard le *Mareschal* of Branketre for 40 shillings of silver, in 1317,FF.

MARTIN CLOSE, Billericay
Christopher *Martin*, a Billericay miller, in 1619 sailed with *The Mayflower* to New England. He died there in 1621.

MARTIN'S ROAD, Halstead
William *Martin*, clothier, died 1573, left provision for the school and a stipend for a master; buried in Halstead church.

MARNEY CLOSE, Chelmsford
Henry *Marney* (died 1523) baron of Layer, Captain of the Guard to Henry VIII, built Layer Marney Towers gatehouse but died before its completion.

MASCALLS WAY, Great Baddow
John *Mascall*, trustee of Society of Friends Meeting House, 1724.

MATCHING HALL, Matching
Robert Gernon held *Matcinga*, Edmund son of Algot held *Matcinge* in 1086,DB. Algot may be a Scandinavian name. *See Matching.*

MAYNARD COURT, Waltham Abbey
John *Maynard*, local historian in the 19th century.

MAYNARD VILLAS, Little Easton
Evelyn 'Daisy' *Maynard* (1861-1938), celebrated beauty, created Easton Lodge gardens on the ancient Bourchier and Maynard estate *(see Warwick Close).*

McINTOSH CLOSE, Romford
Hugh *McIntosh* bought Havering Liberty and the rights of Romford market from the Crown in 1828.

MICHAELSTOW, Ramsey
Ralph Baynard held *Michelestou* in 1086,DB: St Michael's (church) stow, holy place.

MILDMAY ROAD, Chelmsford
Thomas *Mildmay* of Moulsham Hall, Auditor to Elizabeth I, died 1566. *(Fig. 99).*

MILLER'S LANE, Lambourne
Gravel Lane, 1777,C&A.

MILTON HALL, Prittlewell
*c.*1052, King **Edward** confirmed the grant of *Middeltun* to Christ Church Canterbury (ECE 54) which held *Mildentuna* in 1086,DB: tun, middle settlement.

MILTON ROAD, Southend
Mildentuna, 1086,DB: middle settlement between Southchurch and Leigh. Milton Hall is now Nazareth House for the Retired.

MINCHEN ROAd, Harlow
Myncen, a nun. Not far away is Monkswick Road. The Abbey of Bury St. Edmunds held land in Harlow and Latton, and Waltham Abbey in adjacent Netteswell.

MOAT FARM CHASE, Witham
La Mo(o)te, 1370,AD: mot, meeting-place or mote, ditched.

MOLE HILL GREEN, Felsted
Middle English 'mail', land held by payment of rent.

MOLLANDS LANE, South Ockendon
Mollands Farm, Mollande, 13th cent,StJohn: mol land, land held by payment of rent.

MONKHAMS LANE, Woodford
Monekenebukhurst, 1286,AD: Old English munuc bece hyrst, monks' (of Stratford Langthorne Abbey) beech wood.

MONK WOOD, Loughton
Peter de Valognes held *Lochetuna, Lochintuna* in 1086,DB. *Monkwode*, 1498: it belonged to the monks of Stratford Abbey. *See Loughton.*

MONKS AND BARROWS FARM, Writtle
Bedmannesberga, 1176/7; John *Munke* held the estate at Highwood in 1484,TEASxvii,111. It belonged to St. John's Abbey Colchester.

MONKSBURY, Little Hallingbury
Geoffrey Martel granted land of *Halynbury* to Bermondsey Abbey, 1093,(ECE 110).

MONOUX GREEN, Walthamstow
Sir George *Monoux*, draper and Lord Mayor of London, founded Walthamstow Grammar School in 1527.

99. The Thomas Mildmay memorial in Chelmsford cathedral.

MONTBAZON COURT, Brentwood
Brentwood Borough Council is 'twinned' with *Montbazon*, France.

MONTGOMERY CLOSE, Chelmsford
Thomas *Montgomery* of Faulkbourn, lawyer, MP and Keeper of the Mint 1450-66.

MOOR HALL, Harlow
Herlaua was held by Count **Eustace** in 1086,DB: mor, moor. *See Harlow.*

MORANT ROAD, Colchester
Dr. Philip *Morant*, 1700-1770, rector of St. Mary's Colchester and historian. Buried at Aldham.

MORRELL RODING
Geoffrey *Morel* and Joan acknowledged *manerio morelli* to be the right of Alan *Morel* of Geoffrey's gift, 1319,FF. Morrel Roding *alias* Cammas Hall (a detached part of Ongar Hundred) lay north of White Roding and the name continued in use until *c.*1848.

MORLEY ROAD, Halstead
John *Morley* (1656-1735) lived at Blue Bridge House, Halstead. He successfully speculated in East India Stock. Beginning life as a butcher's son he became the agent of the rich.

MORRIS COURT, Chingford and Waltham Abbey
William *Morris* (1834-96), born in Walthamstow, lived at Woodford Hall in Epping Forest. His arts-and-crafts designs for textiles and furniture reflect his rural upbringing.

MOTEHILL, Langdon Hills
Old English (ge)mot, a meeting place.

MOULSHAM, Chelmsford
Caesaromago, 3rd century, Antonine Itinerary: (Celtic) magos, Caesar's market; *c.*1053, (K859), King **Edward** confirmed the grant of *Molesham* by Leofcild to Westminster monastery, which held *Molesham* in 1086,DB: ham, settlement of **Mul** ("half-breed").

MOULSHAM HALL, Great Leighs
Ulmar held *alia Molesham* from **Odo**, bishop of Bayeux, 1086,DB.

MOWDEN, Hatfield Peverel
Mokelinton, 1276,For: ingtun, Mucel's farmstead.
MULBERRY GREEN, Harlow
Mudborow, 16th cent.,LP: gemot beorg, meeting-place (moot) -mound. *See Harlow.*
MUTLOW HILL, Wendens Ambo
Motehlaw, 1316,TEASxvi,183; (ge)mot hlaw, meeting-place mound. The Saxon mot was here.

NASH HALL, High Ongar
John son of Waleran held *Angra* in 1086,DB: anger, grassland. *See also Ongar and Ashlyns.*
NASSAU ROAD, St. Osyth
Frederick *Nassau*, Baron Enfield (1682-1738) acquired St. Osyth Priory through marriage to Bessie, daughter of Earl Rivers. He built Priory House.
NAVARRE GARDENS, Romford
Joan of *Navarre* (died 1437), widow of Henry IV, accused of witchcraft, lived at Havering Palace.
NAYLING ROAD, Braintree
Neilingaherst, 12th cent.,Colne: hyrst, wood of Nægl's people.
NAZE, Walton
Sierra bermeja, *c*.1545 (Spanish navigation chart): red cliffs. *See Walton.*
NELMES ROAD, Hornchurch
In 1213,HPD, Ernwin de *Ulmis* witnessed a lease from Gilbert son of Roger Marschal to Edmund son of Eilmer Prest of land in Havering for 20 marks. *Nelmes*, 1339,Cl: æt elm, at the elms.
NETHER HALL, Roydon
Ranulf brother of Ilger held *Nasinga* in 1086,DB: næss ingas, people of the promontory. *See Roydon. (Fig. 100).*
NEVILLE ROAD, Saffron Walden
Richard *Neville*, Lord Braybrooke (1784-1858) lived at Audley End, bringing valuable portraits and furniture by his marriage to Lady Jane Cornwallis.
NEWCOURT ROAD, Chelmsford
Richard *Newcourt* (died 1716), author of *Repertorium Ecclesiasticum Pariochale Londinense*, the 2nd volume of which details Essex. *(Fig.* ***101****).*
NEWLAND HALL, Roxwell
Count **Eustace** held *Neuuelanda* in 1086,DB: niwe land, new tract of land.

100. The 15th-century gatehouse of Nether Hall, Roydon, in 1817.

NEWLAND STREET, Witham
Newland, 1213,StJohn. The Knights Templar moved the market from Chipping Hill to the London-Colchester Road *c*.1212.

101. Richard Newcourt.

NEWMAN CLOSE, Emerson Park, Hornchurch
Richard *Newman* of Nelmes founded the Essex Hunt, painted by George Romney (1734-1802).

NEWNHAM HALL, Ashdon
Neuuham was held by Count **Eustace** in 1086,DB: niwe ham, new settlement.

NEWTON HALL, Great Dunmow
Niuuetuna was held by **Geoffrey** de Mandeville in 1086,DB: niwe tun, new settlement.

NORSEY WOOD, ROAD, Billericay
Nossesheye, 1250,CottCh: gehæg, Notsye's enclosure.

NORTHEY ISLAND, Maldon
Hamo the Steward held *Carseia* in 1086,DB: eg, island (possibly cærse, marsh-cress, genus Rorippa). *Siricheseie*, 1141,AD: eg, Sigeric's island. *Northeye*, 1413,Bodl: north island.

NORTON (alias Forest Hall), High Ongar
Canons of St. Paul's held *Nortuna* in 1086,DB: north settlement. *Folyats Hall*, *c.*1480,ECP, perhaps from Gilbert *Foliat*, bishop of London. *See also Nash Hall.*

NOAK HILL ROAD, Harold Hill
Noke Hill, 1490,Ct: ætten ac, at the oak.

NORWICH CLOSE, Colchester
Lord George *Norwich* commanded the Royalist defenders of Colchester in 1648.

OCTAVIA DRIVE, Witham
64-11 BC. Daughter of Gaius Octavius, and sister of Emperor Augustus.

OGLETHORPE SCHOOL, Cranham
General James *Ogelthorpe* (1696-1785), sailed to Georgia with new settlers in 1732; died at Cranham Hall.

OLD HEATH, Lexden
Hetha, *c.*1200,Colch; hyth, landing-place. (The Roman hythe was probably further north at the east end of Hythe Street.)

ORGAN LANE, Chingford
Orgersdowne, 1367,TEASii(pt3),196: *Orgar* the thegn held Chingford in 1086,DB; or where *Organa* (pennyroyal, a medicinal and culinary herb) was grown.

ORPEN CLOSE, West Bergholt
George *Orpen*, 20th-century brewer at Daniell's Brewery.

OSBERT ROAD, Witham
King of the Northumbrians, died fighting the Vikings in 867.

OSBORNE ROAD, Dagenham
Hewet *Osborne* of Parslowes, esq., died 1599, traded with Turkey, held land in Clacton, Nottinghamshire, Derbyshire and Yorkshire.

OSEA ISLAND, Great Totham
Hamo the king's Steward held *Uueseia* in 1086,DB: eg, Ufe's island.

OXNEY GREEN, Writtle
Oxenehey, 1274,RH: (ge)haeg, enclosure for oxen.

OYSTER BRIDGE, Great Bardfield
Oyster bridge, 1336,Rental; *Ewestresbregge*, 1363,Ct: eowestre, sheepfold.

PAGE ROAD, Clacton
H.J. *Page* donated land and funds for building of St. Paul's church, opened 1875.

PALINS WAY, Stifford
William *Palin* (1803-82), rector of Stifford, local historian.

PALMER'S AVENUE, Grays
William *Palmer* (1633-1710) son of Edward who purchased the manor of Grays in 1637. He founded Palmer's School.

PALMERS HILL, Epping
Thomas *Palmer* (died *c.*1621), lawyer, lived at Gills.

PAPILLON ROAD, Colchester
Philip Oxenden *Papillon* (1826-99), Conservative MP for Colchester 1859-65.

PARKESTON
Great Eastern Railway's new quay opened 1883; named after its chairman, Charles *Parkes*.

PARNELL CLOSE, Colchester
James *Parnell* (1636-56), Quaker preacher, was imprisoned in the Castle for heresy and died the following year as a result of his ill-treatment.

PASLOW HALL, High Ongar
The canons of Waltham Holy Cross held *Passefelda* in 1086: feld, Passa's open country. *See Ongar*. (Fig. **13**).

PASSINGFORD BRIDGE, Stapleford
Tawney *Pyssingford*, 1224,Wm: (meaning obscure) + ford (through River Roding). *(Fig. 17).*

PATCHING HALL (LANE), Chelmsford
In 1086,DB, **Odo** held the manor of *Pacingas*, settlement of Paecci's people.

PATERNOSTER HILL, Waltham Abbey
Paternoster, 1467,Winters: part of the endowment land for payment of burial of the poor.
PATERNOSTER ROW, Noak Hill
At Rogationtide (about 5 weeks after Easter), the Lord's Prayer was said during Perambulations at points along the parish boundaries.
PATTISON CLOSE, Witham
William Henry *Pattison* (died 1832), Witham shopkeeper and brickbuilder.
PAXMAN AVENUE, Colchester
James Noah *Paxman*, alderman, born at Elmstead, son of a wheelwright, founder of the Colchester engineering works, donated money for building the Colchester Town Hall tower, 1899.
PEMBERTON AVENUE, Fryerning
Rev. Robert (died 1941) and his daughter Miss Vera *Pemberton* (died 1992) who founded Ingatestone Boys' Club in 1919.
PAYCOCKE'S HOUSE, Coggeshall
Thomas *Paycocke* (died 1461) and his son John (died 1505) were butchers; his son Thomas, clothier, (died 1518) and his wife Margaret were clothiers.
PELLY ROAD, West Ham
Sir John *Pelly* (1777-1852), governor of Hudson's Bay Co. (Canada), owned Upton Manor.
PEPYS ROAD, Harwich
Samuel *Pepys* (1633-1703), diarist, represented Harwich as MP in 1679 and 1685.
PERRY STREET, Billericay
Perystrete, 1542,Ass: pirige, pear-tree.
PETRE CLOSE, Ingatestone
Sir William *Petre* (1505-1572) of Ingatestone Hall, was Secretary to Henry VIII. His descendants continue to live at the Hall. *See Tor Bryan, and Wadham Chase.*
PETTITS LANE, Romford
*c.*1235,HPD, Adam de Lincoln was given right of way through the court and fields of Thomas *Petit*, for 1d yearly. Great Pettits was *Petites*, 1478,Ipm.
PEYTON HALL, Manuden
Pachen hov was held by William de **Warenne**, 1086,DB: Paca's hoh, promontory.
PHILPOT END, Great Dunmow
Filberd End, 1777,C&A: associated with Thomas *Fylpote*, 1544,ECP.
PICK HILL, Waltham Abbey
pic, conical hill above Cobbins Brook.
PICKET'S BUSHES, East Horndon
William *Piket*, relative of Robert *Pycot*, draper of London, 1365/6,FF,TEASxvii,109.
PILGRIMS HATCH, Brentwood
Pylgremeshacch, 1483,Will: Middle English pelegrim hacche, forest-gate into Brentwood for pilgrims visiting the chapel dedicated in 1221 to St. Thomas (now in Brentwood High Street) on their way to Canterbury.
PINCHPOOLS, Manuden
Sasselin held *Pincepo* in 1086,DB: possibly pinc or pinca pol, minnow or finch pool (on R. Stort).
PLAISTOW, West Ham
Playstowe, 1414,Pat: plegstow, a place for playing.
PLEGDON HALL, Henham
Eudo the Steward held *Plicedana* in 1086,DB: Plycca's? denu, valley.
PINFOLDS, Aveley
Middle English pinfold, enclosure or pound.
PLASHET ROAD, East Ham; (PLASHET GARDENS, Ingrave)
John atte *Plasshet*, 1327,SR. Plashet hamlet takes its name from Norman-French 'plessis', that part of Epping Forest enclosed by a living hedge fence. Elizabeth Fry (1780-1846) lived at Plashet House, East Ham, 1808-29.
PLUMBEROW MOUNT, Hockley
Plu'bga was held by William de **Warenne** in 1086,DB: plume beorg, plum-tree mound. 'Mount' is a tautology. There is a tumulus, probably Roman, at TQ839938.
PLUME AVENUE, Maldon
Thomas *Plume* (1630-1704), Archdeacon of Rochester, endowed a chair of Astronomy at Cambridge and left his library to Maldon where he was born. *(Fig. 102).*
PLUNKER'S GREEN, Doddinghurst
Ranulf Peverel held *Plumtuna* in 1086,DB: plume tun, plum-tree settlement. *See Doddinghurst.*
POCKLINGTON CLOSE, Chelmer Village
Rev. Oliver *Pocklington*, rector of Chelmsford for 33 years from 1706. His relative Thomas became a Trustee of the Charity School, founded 1713.

102. Thomas Plume of Maldon.

PODS LANE, Rayne
No doubt takes its name from *Pod's Brook*, 1777,C&A: pode, a toad, or a person called Pod.

POINT CLEAR, St Osyth
Apparently Stone Point, with reference to St. Clere's Hall (Cicely St. Clere, 1273, descendant of Lord of St. Cler, France, who aided **William I** in 1066).

POINTWELL LANE, Coggeshall
Poyntelmelne, 1261,FF: Theodoric Pointel ('the sharp one') held Coggeshall in 1086,DB.

POLE HILL ROAD, Chingford
The canons of St. *Paul's* held the manor of Chingford in 1086,DB.

POOLEY (*alias* Hunts Hall), Pebmarsh
William de **Warenne** held *Polheia* in 1086,DB: pol (ge)haeg, enclosure by a pool. Ranulf de *Polhey* acknowledged the right of Simon le *Hunte* to 1 messuage, 1 mill, 4 acres meadow etc. in Pebeners, 1312,FF.

POORS LANE, Hadleigh
Land from which the parish poor were supported.

POOTY POOLS, Roxwell
In 1206,FF, Thomas fitz Abraham, tenant, quitclaimed all that land held by Richard de Pudipol in *Pudipol*: O.E. ?pudde pol, ditch pool.

PORT LANE, Hallingbury
Middle English port, town/market, but here probably harbour gate to Great Hallingbury.

PORTWAY COURT, Halstead
Charles *Portway* & Son factory manufactured cast mouldings for machinery from 1870s onwards.

POTTER STREET, Harlow
Pottereshill, 1254,Ass: Cok le *Pottere* lived here in 1291,For. In Potterestrate, 1382, Maurice the *potter* held land.

POTTER STREET, Sible Hedingham
In 1444,FF, John Vampage held Hawkwodes of *Potterstrete* from Thomas Broun and Thomas Payne.

POWDERMILL WAY, Waltham Abbey
Site of the mills for making gunpowder and explosives since the 1600s until 1991. *(Fig. 103).*

POWERS HALL END, Witham
Robert Gernon held Witham with a mill in 1086,DB; Robert le *Poher*, 1250,CottCh.

PRESTED HALl, Feering
Ranulf Peverel held *Peresteda* in 1086,DB: peru stede, pear place.

PRIORY STREET, Earls Colne
The priory was founded by **Aubrey de Vere** in 1101-7. After its dissolution in 1563 it passed to the Earls of Oxford.

PUCK LANE, Waltham Cross
Puchisland, 1356,Winters, was held by Robert de Pukelond: puca, goblin.

PUDSEY HALL, Canewdon
Puteseia was held by **Swein** in 1086: hyth, Pud's landing-place.

PURFLEET, Thurrock
Purteflyete, 1285,StJohn: fleot, inlet/stream. First element may be a personal name *Purta*. Purfleet is at the issue of Mar Dyke to the Thames. *(Fig. 104).*

103. The gunpowder mills, Waltham Abbey.

PUTWELL BRIDGE, near A12/M25 junction
A12 crosses Ingrebourne river at *Delbregge*, 1276,For; Leticia *Potehale*, 1272,Ass; John *Putehale* of Upminster quitclaimed 1 messuage and 8 acres in 1374,FF.

PYEFLEET CHANNEL, Mersey
Pyefletemersch, 1378,Ct: pye, oyster-catcher; fleot, inlet, mersc, marsh.

PYEFLEET CLOSE, Brightlingsea and Fingringhoe
Pye, oyster-catcher + fleot, inlet.

PYRLES LANE, Loughton
Pirleswente, 1404,Loughton. John *Pyrle* held land by Loughton Bridge *c.*1200.

PYNCHON MEWS, Springfield
John *Pynchon* (1626-1703), born at Springfield, sailed for New England in 1630. Becoming a successful fur-trader he was elected Treasurer of Springfield, Massachusetts.

QUARLES CLOSE, Romford
Francis *Quarles* (1592-1644) poet and chronologer to City of London, born at Stewarts, Romford. Later settled at Roxwell. Died of consumption and buried at St. Vedast's, London.

QUICKSBURY, Sheering
Cuica was held by William de **Warenne** in 1086,DB: cu wic, cow dairy-farm.

RAINHAM HALL, Rainham
Walter of Douai (Nord) held *Reineham* in 1086,DB. *See Rainham.*

RAMPART STREET, Shoeburyness
refers to the D-shaped earthwork marking the enclosure possibly of Viking origin.

RAMSEY ROAD, Halstead
Dame Mary *Ramsey*, wife of Sir Thomas, Lord Mayor of London, founded the Grammar School in 1594.

RANULF WAY, Hatfield Peverel
Ranulf Peverel, Domesday holder of extensive estates in Essex, died *c.*1090. *See Ingelrica Avenue.*

RAPHAEL AVENUE, Gidea Park
Sir Herbert *Raphael*, MP, gave Raphael Park to the town *c.*1935.

104. Purfleet in 1807.

105. Valentines House, Ilford.

RAYMOND ROAD, Ilford
Sir Charles *Raymond* lived at Valentines, 1769. *(Fig. 105).*

REBOW ROAD, Wivenhoe; - STREET, Colchester
John *Rebow* purchased Colchester Castle in 1685. His son Isaac, MP, was knighted. His grandson Isaac of Wivenhoe Park married Mary, daughter of Captain Martin of Wivenhoe.

REPTON GARDENS, Romford
Humphry *Repton* (1752-1818) lived at Balgores Lane, Romford. He gained a reputation for landscaping estates including Rivenhall Place, Hill Hall, Theydon and Woodford Hall. *(Fig. 106).*

RICH CLOSE, Felsted
Sir Richard *Rich*, ?1496-1567, first baron, MP for Colchester 1529, suppressor of monasteries for Henry VIII, Lord Chancellor 1548, founder of Felsted School in 1564, was buried in Felsted church. *(Fig. 36).*

RIDLEY ROAD, Chelmsford
Thomas *Ridley*, Chelmer miller, later the firm T.D. Ridley, brewers and maltsters, incorporated 1906.

RIDLEY HALL, Terling
Geoffrey de Mandeville held *Retleia* in 1086: read leah, red wood?

RIGBY AVENUE, Mistley
Richard *Rigby* (died 1730) successfully invested in South Sea Stock, rebuilding Mistley Hall.

RISEBRIDGE ROAD, Romford
Risbregg, 1230,HPD: hris brygg, causeway reinforced with brushwood.

ROBINSON CLOSE, Hornchurch
The pilot Capt. William Leefe *Robinson*, VC, shot down the first Zeppelin in 1916.

ROCKELLS, Elmdon
Formerly *Wigghepet* which was annexed by William de **Warenne** in 1086,DB: Wicga's pyt, hollow. *Unfrey de Rokesle* held from Roger de Nufvill 2½ virgates in Anmodone, 1205,FF.

106. Repton's 'View from his cottage of Hare Street' (Main Road), Romford, 1816.

ROLPHY GREEN, Great Waltham
Rocheye, 1250,CottCh: ruh (ge)haeg, rough enclosure. See Waltham, Great.

ROMFORD, Main Road (A118)
Documented in 1344,Ct, as *Herstrate*, from Old English here-stræt, army-path. *(Fig 106)*

ROTH DRIVE, Brentwood
Brentwood Borough Council is 'twinned' with *Roth* in Germany.

ROWHEDGE, East Donyland
Rouhegy, 1346,Ct: ruh (ge)haeg, rough enclosure.

ROWLEY LANE, Kelvedon
Rawley, 1545,Ct: ruh leah, rough clearing in a wood.

RUCKHOLT, Leyton
Leintuna was held by **Peter** of Valognes in 1086,DB. Robert fitzwalter gave 5 acres of meadow next to *Rocholt* to Celemencia prioress of Haliwelle in Luiton, 1200,FF: hroc holt, rook wood. *See Leyton.*

RUSH LANE, Elsenham
George *Rush* lived at Elsenham Hall in early 19th century.

RYES, Hatfield Broad Oak
Siriceslea was held by Hamo the Steward in 1086,DB: Sigeric's leah, wood/clearing. *Ryse Morses*, *c.*1540,ECP: hris, brushwood, Ralph de Marci was undertenant in 1086.

SABINE'S GREEN, Navestock
Grant of Adam fitz William to Hamund de Berga 8 acres in Havering between Pirifeld and land of William fitz *Sabine*, *c.*1235,HPD.

ST. ANDREWS DRIVE, Billericay
Here, until 1998, was the Plastic Surgery Unit, St. Andrews Hospital named *c.*1935, formerly Billericay Union Workhouse founded 1838. *(See Burns Close). (Fig. 78).*

ST. ANN'S CASTLE, Great Leighs
St ***Ann's*** *Ermitage*, 1571,M. A hermitage hospice for pilgrims stood hereabouts before the Dissolution. It was an inn by 1636.

ST. CHAD'S ROAD, Tilbury and Romford
Chad, younger brother of St. **Cedd** who founded a church at Tilbury, was trained as a priest at Lindisfarne, became bishop of York and Lichfield, and died of plague in 672.

ST. CLERE'S HALL, Danbury
Sanclees, 1412,FA; John de Sancto Claro, *c.*1277,TEASxvii,106. *(See Point Clear, St. Osyth).*

ST. ERKENWALD ROAD, Barking
Erkenwald, co-founder with his sister Ethelberga of Barking monastery *c.*666. Probably a member of the royal family, he was bishop to the East Saxons at St. Paul's. Died ?693.

SAMPSONS, Peldon
Samantuna was held by **Odo** in 1086,DB: Sæmann's tun, settlement. *See Peldon.*

SANDERSON MEWS, Colchester
Richard *Sanderson* represented Colchester as MP 1833-1845.

SAULS BRIDGE CLOSE, Witham
Sallys Bredge, 1549,Pat, (over the River Brain) associated with John *Salle*, 1381,SR.

SCOTTS HILL, Southminster
Alexander *Scott* (1768-1840), vicar of Southminster, naval chaplain. Administered last rites to Nelson as he lay dying on *HMS Victory* at Trafalgar, 1805.

SEVEN KINGS, Ilford
Sevekyngg, 1285,Ass: settlement of Seofeca's people or meeting-place of seven Saxon kings.

SEVERALLS LANE, Colchester
Several Hall, 1777,C&A: severals, lands in private ownership.

SEWARDSTONE (ROAD), Waltham Abbey
Siwardeston', 1176-90, P: Sigeweard's settlement. *See Waltham.*

SEWARDSTONE STREET, Waltham Abbey
Formerly *Eldewirche, c.*1170,P: eald worth, old enclosure. *See Sewardstone.*

SHEDDON (OLD) HALL, Mistley
Sciddinchov was held by the Countess of Aumâle (half-sister of William I), 1086,DB: scydd inga hoh, shed-dwellers' hill. *See Mistley.*

SHEDS LANE, Saffron Walden
Thomas *Shede* held a tenement in 1564,HMC.

SHEEPEN ROAD, Colchester
Shepynlane, 1550,Pat: scipen, cow shed.

SHELLEY GROVE, Loughton
Thomas son of John *Shelley,* 1567,SR, who held Roukes.

SHEEPCOT ROAD, Castle Hedingham
Old English: sceap cot, outbuilding for sheep.

SHELL HAVEN, Corringham
Shell Haven, 16th century,Cott: ?scylf hafn, bank harbour. Coincidentally Shell Oil Company opened a refinery here in 1912.

SHILLIBEER WALK, Chigwell Row
George *Shillibeer* (1797-1866) lived at The Grove. In 1829 he introduced and drove the first London omnibus along the Marylebone Road to the City of London.

SHIR BURN, Lawford
Old English scir, boundary, between Tendring and Lexden Hundreds?, or scir burna, gleaming stream.

SHOELAND, Colchester
William II granted *Sollanda* to **Eudo** Dapifer, *c.*1099,Colch. 'Solanda' was a desmesne estate held by a cathedral canon to provide footwear from cattle hide. (Site now is at the Grammar School, TL984247.)

SILVER END, Rivenhall
Silver End, 1834; William **Siveley** married Elizabeth Green in 1734: EAH News144,p5.

SILVERLOCKE ROAd, Grays
James *Silverlock* lived at Ford Place, Stifford, 1655.

SIR ISAAC'S WALK, Colchester
Isaac *Rebow,* MP for Colchester *c.*1700. *See Rebow Road.*

SIWARD ROAD, Witham
Siward of Maldon, Saxon thegn, joined Hereward the Wake in opposition to William the Conqueror in The Fens, 1070-71.

SKIGHAUGH, Great Oakley
John Flinthard held 80 acres in Great Acle from Peter de *Skeghawe* for 10 marks, 1256,FF: ?Scandinavian personal name + Old English haga, Skegge's enclosure.

SKREENS, Roxwell
William *Skren,* serjeant-at-law at Clifford's Temple, London, and Richard Gybbon quitclaimed from John Rote a messuage, 60 acres of land, 4 acres of meadow and 4 acres of wood in Wretylle in 1389,FF. The last of the line was Sir John *Skrene,* died 1474, who was knighted after the battle of Tewkesbury. *(Fig. 107).*

SLADE, THE (Saffron Walden)
Old English slaed, valley.

SLAMSEYS, Black Notley
Bishop of London, held *Slamondesheia* in 1086,DB: sla munts (ge)haeg, sloetree hill enclosure.

SLOW HILL, Halstead
Site of *Slohouse,* 1412; a tenement *Slowhouse* in tenure of George Sexten, 1548, TEASxiv,332: Old English sloh, mire.

SMALLGAINS, Stock
A derisory name for an unproductive field.

SMALLLANDS (Marshalls), Hatfield Peverel
Robert son of Corbucion held *Smalelant* in 1086,DB: smæl land, narrow tract of land. Thomas *Marscal,* 1355,FF. Christofer Edmonds to Thomas Wysman the manor of *Smallondhall als Marsshe,* with 2 messuages, 200 acres etc. in Hatfeld Peverell, Wytham and Fayrested, 200 marks, 1544,FF.

107. Skreens and Shellow Bowells sign.

SMALLTON, Castle Hedingham
Aubrey de Vere annexed *Smaltuna* in 1086,DB: smæl tun, narrow settlement. *See Hedingham.*

SMART'S LANE, Loughton
Mathew *Smart* held a cottage and slip in Allards als *Smarts* Lane in 1819,Loughton.

SMEETHAM, Bulmer
Count **Eustace** held *Smedetuna* in 1086,DB: smethe tun, level settlement. It lies on a plateau above River Stour.

SNARESBROOK, Wanstead
Snaresbrook, 1599,TEASvii: snor, twisting, or snar, brushwood + brook.

SNAKES LANE, Woodford
Sake's Lane, 1583. In 1404,ENvii,p25, John *Sake* of Motteslane and his son, brewers, the latter being fined for not keeping his ditches scoured.

SORREL CLOSE, Little Waltham
Richard *Sorrel* (died 1738) is commemorated in St. Mary and St. Lawrence church.

SOUTHALL (*alias* CLOPTON), Great Dunmow
William de **Warenne** held *Dommauua* in 1086,DB. William earl of **Bohun** and Northampton and Elizabeth held the manor and advowson of Neuton which William de *Clopton* held for his life, 1345,FF. *See Dunmow.*

SOUTHCOTE ROAD, Witham
John *Southcote* (1511-85), judge, bought Bacons and Abbots Witham *c.*1565. His grandson Sir Edward (1658-1751) was imprisoned by the Parliamentarians after the Battle of Newbury in 1643.

SPAINS HALL, Willingale Spain
Ulingehala was held by Count **Alan** in 1086,DB: *See Willingale.*

SPALDING AVENUE & WAY, Chelmsford
Fred *Spalding* moved from Danbury to Chelmsford *c.*1859, trading as a bird-stuffer and later photographer. His photographs are a valuable record of 19th and early 20th century life.

SPANSEY COURT, Halstead
Spermesyspringe, 1548,TEASxiv: spring, a small wood.

SPAYNES HALL, Great Yeldham
Count **Alan** held *Gerham* in 1086,DB. John de Ispania (Épagnes, Eure), 1285,Ass. *See Yeldham.*

SPITAL LANE, Brentwood
Here was a leper hospital, St. Giles, in the 13th century.

SPITAL ROAD, Maldon
A leper hospital here was founded by Henry II in 1164.

SPURGATE, Hutton, Shenfield
Entrance to a right-of-way for a horse-rider, a spurway.

SPURGEON PLACE, Kelvedon
Charles *Spurgeon* (1834-92), born at Kelvedon, attracted a great following due to his sermons. His Baptist Hall at Newington, London held a congregation of 6000. *(Fig. 108).*

SQUADRONS APPROACH, Hornchurch
RAF Hornchurch operated from this airfield in 2nd World War. *See Airfield Way.*

SQUIRRELS HEATH, Romford
Squyrells, 1525,Husting: Geoffrey *Scurell* was witness to grant of meadow in Havering Marsh of Laurence del Brok to Richard of Havering *c.*1261,HPD.

STANE STREET, Colchester-Bishop's Stortford (A120)
Stanstrete, 1181,StP: OE stan stræt, paved road, probably Roman or earlier.

STANESGATE PRIORY, Steeple
Stanesgata was held by **Ranulf** Peverel in 1086,DB: gatu, Stan's gates (is owned by the Wedgwood Benn family).

STANSTEAD HALL, ROAD, Halstead
Robert Malet held *Stanesteda* in 1086,DB: stan stede, stoney place.

STANWELL STREET, Colchester
Old English stan wielle: led to the stone-lined well.

STEBBINGFORD, Stebbing
Adam son of Durand Malzor ('evil deeds') held *Horstedafort* in 1086: horh stede ford, mire place ford.

STEVINGTON END, Bartlow
Stauintuna was held by Count **Alan** in 1086,DB: tun, Styfa's settlement.

108. Charles Spurgeon, renowned preacher.

STICKLING GREEN, Clavering
Stiklyngrene, 1421,MnAt: Stephen *Stikeling* rented 15 acres of land in 1258,FF.

STOCKING GREEN, Radwinter
Stocking Green, 1777,C&A: stoccing, land cleared of tree-stumps.

STOCKWELL STREET, Colchester
Old English, stocc wielle: 'the spring by the tree-trunk'.

STONARDS HILL, Loughton
John *Stonard* (died 1540), yeoman, was a tenant of Loughton manor.

STOURTON ROAD, Witham
William *Stourton,* Speaker of House of Commons in 1413.

STUBBERS LANE, North Ockendon
William *Stubber,* yeoman farmer, lived at *Stubbers* *c.*1419.

STUDLEY WOOD, Woodham Ferrers
Hugh de Montfort held *Estoleia* in 1086,DB: stod leah, horse-stud clearing. *See Woodham.*

STRANGMAN AVENUE, Hadleigh
William *Strangman* esq., died 1573, held lands in Purleigh, Rawreth and Rayleigh. Buried at Hadleigh church.

STRETLEY GREEN, Littlebury
In 1008, (K725), King **Æthelred** sold *Strethle* for £9 in gold from Ælfsige, abbot, to Ely monastery: stræt leah, street wood/clearing?. *See Littlebury.*

STRUTT CLOSE, Hatfield Peverel
John *Strutt* (1842-1919), 3rd Baron Rayleigh, FRS, Cavendish professor of Physics at Cambridge, retired to Terling Place.

STUKELEY ROAD, East Ham
Dr. Wiliam *Stukeley* (1687-1765), FRS, antiquary, is buried in St. Mary Magdalene church.

SURMAN CRESCENT, Hutton
Secretary of the South Sea Company founded 1711, speculation in which led to economic and monetary collapse ("South Sea Bubble") in 1720, lived at Hutton Hall.

SUTTONS AVENUE, Lane, Hornchurch
Sutton, 1257,HPD. Site of Sutton Farm on which Hornchurch Airfield was built. *See Airfield Way.*

SWEYN AVENUE, Hawkwell
Swein (died *c.*1086), Sheriff of Essex under **Edward** the Confessor and **William I**, held land in Hockley, Rayleigh, Sutton, Clavering etc.

TABORS AVENUE, Great Baddow
George *Tabor,* a tailor in Baddow Road, *c.*1884.

TANNER STREET, Barking
Street where tanners and leather-workers predominated.

TAYLOR AVENUE, Braintree
John *Taylor* and Edward English signed the Trust Deed of the first Congregational Church in Braintree in 1709.

TAYSPILL COURT, Colchester
The 17th-century Flemish *Tayspill* family brought prosperity through the former weaving trade.

TEMPLAR ROAD, Cressing
Knights *Templar,* soldier knights to the Crusades formed *c.*1180; their many possessions were stripped from them in 1308 by Edward II and some transferred to the Knights' Hospitaller. *(Fig. 109).*

TENNYSON AVENUE, Waltham Abbey
Alfred Baron *Tennyson* (1809-92), poet laureate, lived at High Beach. His *In Memoriam* (1850) was inspired by the Waltham Abbey bells.

109. *The Templar 13th-century wheat barn, Cressing.*

TENPENNY HEATH, Thorrington

Turnpeny hamlet, 1594,N: associated with John Turnepany, 1337,Ct.

TENSING GARDENS, Billericay

Sherpa *Tensing* Norgay and Edmund Hillary reached the summit of Mt. Everest on 29th May 1953.

TENTERFIELDS, Wickford, Nevendon and Newport

After fulling, cloth was stretched (*tented*) on frames to dry. These roads occupy former cloth-workings.

THAMESHAVEN, Coryton

Thames Haven Dock was served by a railway connecting to the Tilbury-London line, 1875.

THOBY LANE, Mountnessing

William and Matilda Murmium granted to the prior of *Ginges Thobye* 1 mark yearly rent to be received in Cupeland, 1242,FF: ginge, people of the district. Tobias was the first prior of the priory *c*.1150. *(Fig. 110).*

THOMPSON CLOSE, Ilford

Eleanor *Thompson*, died 1878, of Clements, built the Infants' School and Reading Rooms.

THORPE BAY, -HALL AVENUE, Southchurch

Ranulf brother of Ilger held *Thorp* in 1086,DB: Old English or Old Scandinavian throp, thorp, settlement.

THREADNEEDLE STREET, Chelmsford

Walter Brochole owned a freehold messuage in *Brocholstrate*, 1344; *Brochelelane*, *c*.1590,TEASxix,39. O.E. brocc-hol=badger sett.

THREMHALL PRIORY, Stansted Mountfitchet

Tramhale, 1194,BL: healh, Thrymma's corner.

THUNDERLEY, Wimbish

Tunreslea was held by **Aubrey** de Vere in 1086,DB: Thunor's leah, clearing. *(Fig. 33).*

THURSTAN ROAD, Witham

Saxon thegn, son of **Lustwine**, who left land in Harlow, Pentlow and Ashdon to his wife Æthelgyth and Bury St. Edmunds monastery, *c*.1044 (ECE 59).

TIBERIUS GARDENS, Witham

Second Roman emperor, AD14-37. Appointed Pontius Pilate governor of Judea.

TILKEY ROAD, Great Coggeshall

Tylkell, 1566,Ct: Middle English tile-kelne, tile kiln.

TILLINGHAM HALL, Childerditch

Ciltedic was held by **Swein** in 1086,DB. Richard de Tillingham held it, 1165,M. *See Childerditch.*

110. *The remains of the 14th-century Thoby Priory church, Mountnessing.*

111. Sir Nicholas Coryngham Tindal statue, Chelmsford.

TINDAL SQUARE, Chelmsford
Sir Nicholas Conyngham *Tindal* (1776-1846), Chief Justice of Common Pleas. Lived at Coval Hall, *qv. (Fig. 111).*

TOMLYNS CLOSE, Hutton
John *Thomelyn*, 1472,Ct: *Tomlyns* House was the home of Lilias Rider Haggard.

TOMSWOOD HILL, Ilford
Thom at woodes hill, 1590,Ct; Tomswood (House), 1777,C&A; William *Tom* witnessed the acknowledgement by Geoffrey fitz Peter of Rumford of a debt of 103s 8d to Imbert Pergeis, 1245,HPD.

TOOT HILL, Stanford Rivers
Toote Hill, 1611,Rental: tot, look-out place at 103m OD. (*Essex Journal*, 2007, vol 42, p. 11).

TOR BRYAN, Ingatestone
Named from the hamlet in Devon, seat of the Petre family, 1539. *See Petre Close.*

TOT LANE, Birchanger
Possibly Old English tot, a look-out place.

TRAPS HILL, Loughton
Radulf als Ralph *Traps*, tenant of Tupedene (i.e. Debden Hall, qv) paid ½d wardpenny in 1236,Loughton. John *Traps*, offender, son of William of Schadeshill, 1400,Loughton.

TREWS GARDENS, Kelvedon
Trewes, 1620,HMCx: associated with John *Trewe*, 1417,AddCh.

TRIMBLE CLOSE, Fryerning
Rev. Claude S. *Trimble*, rector of Fryerning, 1927-57.

TRUELOVES LANE, Ingatestone
Thomas *Trywelove*, c.1275,Ingatestone Domesday; John *Trewlove* held land in Ingatestone, 1382,PetreArchives.

TRYON COURT, Halstead
Sir Samuel *Tryon* (died 1626), merchant, rebuilt Boys Hall.

TUFNELL WAY, Colchester
Samuel *Tufnell* of Langleys, Waltham, MP for Colchester in 1727.

TUMBLER'S GREEN, Stisted
Stumblytye, 1382,Ct: Old English stumbel teag, tree-stump enclosure.

TURKSHILL WOOD, Buttsbury
Walter *Turk*, citizen of Olde Fisschstret, London, keeper of the king's exchanges, held a wood in Buttsbury, before 1340,TEASxvii,107.

TURPIN'S LANE, Woodford
Richard *Turpin* (1705-39) began waylaying travellers through Epping Forest. He was hanged at York.

TUSSER CLOSE, Rivenhall
Thomas *Tusser* (1524-80) born at Rivenhall, proverbist, poet and musician at the royal court.

TWITTY FEE, Little Baddow
Tewdy Fee Wood, 1777,C&A: Old English feoh, possession; William Twitye, 1548,Harl.

TWIZZLEFOOT BRIDGE, Burnham
Twisleworth marsh, 1539,Harl: twisla werth, forked-stream/marsh. Here in the marshland east of the village several forking streams drain southwards.

TYLNEY ROAD, Wanstead
Richard Child, later earl *Tylney* of Castlemaine, built Wanstead House in 1715. *See Wellesley Road. (Fig. 67).*

TYRELLS CLOSE, Chelmsford
Sir Henry *Tyrell*, first governor of the grammar school, 1551.

ULEHAM'S, Latchingdon
Bishop of London held *Uleham* in 1086,DB: ule ham, owl settlement.

UNWIN PLACE, Stock
William *Unwin* (*c*.1745-86), rector of Stock in 1769, was acquainted with William Cowper, the poet, who wrote 'Tithing Time at Stock' after staying with him.

UPSHIRE ROAD, Waltham Abbey
Upescire, 1274,RH: Old English upp scir, upper district.

URSWICK GARDENS Dagenham
Sir Thomas *Urswyck* (died 1479) MP for City of London, Chief Baron of the Exchequer, lived at Marks, Romford.

VALENCE ROAD, Dagenham
Agnes de *Valencia*, 1309,Ipm: Aymer de Valence, earl of Pembroke (died 1324 in Paris) inherited the manor in 1309. Valence House is mainly 17th century on a medieval moated site. *(Fig. 112)*.

VIA DEVANA
Name concocted by Dr. Mason of Cambridge in 18th century for the Roman road between Colchester-Lexden-Sible Hedingham-Pool Street-Great Yeldham-Cambridge-Chester (Latin Deva).

VICTORIA ROAD, Chelmsford
Before 1856, *School Lane*, site of the National School.

WADHAM CHASE, Ingatestone
Dorothy *Wadham* (1534-1618), daughter of Sir William Petre of Ingatestone, founded Wadham College, Oxford, 1612. *See Tor Bryan.*

WALDEGRAVES, The, Bures
Waldegrave family bought Smallbridge Hall, Bures St. Mary in 14th century. Lord Henry (1659-1689) was Comptroller to the Royal Household of King James II.

WALKER AVENUE, Fyfield
Anthony *Walker*, rector of Fyfield 1662-1692, endowed the school.

WALLASEA ISLAND, River Crouch
Walfliet, 13th cent,BL: weall fleot, sea-walled inlet/estuary.

WALLENGER AVENUE, Romford
J.A. *Wallenger* lived at Hare Hall, Romford, *c*.1768.

112. *Valence House, from the* History of Dagenham *by J.P. Shawcross, 1904.*

WALSINGHAM ROAD, Colchester
Sir Francis *Walsingham*, (*c.*1530-1590), by espionage obtained plans of the Spanish Armada, Recorder for Colchester in 1583.

WALTER HALL, Boreham
Walhfare, 1062, (K813), was one of **Harold's** gifts to Waltham Holy Cross. The canons held *Walcfare* in 1086,DB: Welshmen's passage (a reference to the native Britons). The Hall was at *c.*TL736108 and the manor reached from woodland in the north to River Chelmer. The 'passage' was the road from Chelmsford to Colchester.

WALTON GARDENS, Waltham Abbey
John *Walton* owned the Gunpowder Mills in early 18th century. *(Fig. 103)*

WANTZ LANE, Rainham
wente, path/road.

WARNERS CLOSE, Woodford
Richard *Warner* (1711-75), lawyer, lived at Woodford Green. He wrote on botany and kept an exotic garden at Harts House here.

WARWICK CLOSE, Little Easton
Francis Evelyn (Daisy) Maynard, wife of Lord Brooke Greville, 5th earl of *Warwick*, in the circle of the Prince of Wales, lived at Easton Lodge *(see Maynard Villas)*.

WASHINGTON CLOSE, Maldon
Laurence *Washington* (1602-52), rector of Purleigh, great-great-grandfather of George, first President of America. The family shield was 3 stars and 2 stripes. Buried at All Saints, Maldon.

WATERLOO LANE, Chelmsford
Shytburye Lane, leading to Springfield Mill, *c.*1590,Grieve: scytel, dung/sewer.

WATERMANS END, Little Laver
Waterman's-end, 1768,M: John *Waterman* was a witness to a Charter of lands of John Bataylle,1398,Cl.

WATLING LANE, Thaxted
Walkyngstrete, 1461,MnAt: John *Walkelyn* held 2 burgage plots rendering 14d per annum in 1348,Cl.

WATTON'S GREEN, Navestock
Joan widow of John de Hobregge quitclaimed a messuage, 30 acres in Navestock of John and Margaret de *Walton* in 1319,FF.

WATTS YARD, Manuden
James *Watt*, fellmonger and glover, had a leather business here in the 18th century.

WEDLAKE CLOSE, Hornchurch
Two brothers *Wedlake* started the Fairkytes Foundry in Billet Lane in 1784 making farm implements.

WELL FARM, Great Burstead
Bishop of London held *Wella* in 1086,DB: wielle, spring/stream. *See Burstead, Great.*

WELLESLEY ROAD, Wanstead
William *Wellesley*, Earl of Mornington (1788-1857), Lord Warden of Epping Forest, of Wanstead House and Park gambled himself into poverty, the House being demolished after its sale in 1822. *See Tylney Road.*

WELLINDITCH, Stow Maries
? *Weneswic* was held by Geoffrey de **Mandeville** in 1086,DB: ?wesung wic, soaking farm ?flax working.

WESLEY AVENUE, Colchester
John *Wesley* visited the Methodist church in Colchester in 1772.

WESTERN ROAD, Rivenhall
Thomas *Western* (died 1692) of Rivenhall Place, merchant. Charles Callis, Baron *Western* of Felix Hall (1767-1844) MP, antiquarian collector and agricultural reformist.

WESTEND HALL, Foxearth
Roger Bigot held *Westuna* in 1086,DB: west tun, settlement.

WHARF ROAD, (and NAVIGATION ROAD), Chelmsford
In 1797 was opened the Chelmer & Blackwater Navigation (Canal) between Chelmsford and Heybridge, Maldon.

WHEATLEY, Rayleigh
Swein held *Watelea*, 1086,DB. Abbot G[ilbert] of Westminster monastery rented to Robert son of **Swein** *Wateleya* for 60s per annum, *c.*1087,(ECE 99): hwæte leah, wheat-clearing.

WHIPPS CROSS, Ilford
Phypps Cross, 1517,WW. John *Phippe* and others held 2 knights' fees in Cardons, Chishull in 1374,Cl. The Hospital, formerly West Ham Union Infirmary, was built around 18th-century Forest House.

WHITMANS (GREAT), Purleigh
Robert Gernon held *Witham* in 1086,DB: Hwita's ham, settlement. *See Purleigh.*

WHITMORE CLOSE, Orsett
Sir Francis *Whitmore* (1872-1962), Lord Lieutenant of Essex and Lord of the manor of Orsett, commanded the Essex Yeomanry in the 1st World War.

WIGLEY BUSH LANE, Brentwood
Wygleye,1403,Ct: weoh leah, pagan shrine clearing ?

WIGNALL STREET, Lawford
Wyggenhale, 1255,Ass: healh, Wicga's corner/raised land.

WIGRAM SQUARE, Walthamstow
Robert *Wigram* (died 1830), of Walthamstow House, ship's surgeon to the East India Co., high sheriff of Essex.

WILBYE CLOSE and HOUSE, Colchester
John *Wilbye* (*c*.1573-1638), born at Diss, composer of madrigals, lived in Trinity Street, Colchester.

WILLINGALE CLOSE, Loughton
In 1860s, Thomas *Willingale* had 'the right to lop trees'.

WILMER LEA CLOSE, West Ham
George *Wilmer* esq., died 1594, held a house in Aldersgate Street, London, and in Southwark.

WILMOT GREEN, Warley
Miss Ellen *Wilmot* (1858-1934) of Warley Place, horticulturist and musician.

WIMARC CRESCENT, Rayleigh
In 1086, **Swein**, grandson of *Wymarc* (a Breton woman), held Rayleigh; he built the castle.

WINSLEY ROAD, Square, Colchester
Arthur *Winsley* (died *c*.1727) endowed almshouses, Old Heath Road.

WINSTANLEY ROAD, Saffron Walden
Henry *Winstanley* (1644-1703), artist and engineer, born in Saffron Walden, clerk of works at Audley End, lived at Littlebury, designed the first Eddystone lighthouse southwest of Plymouth, drowning there during a storm.

WINSTREE ROAD, Beacon End, Colchester
Wensistreu, 1086,DB; treow, Wynsige's tree.

WINTERBOURNES, Walton-on-Naze
Named from the stream which flowed only in winter.

WINTERS WAY, Waltham Abbey
William *Winters* (1834-93), book-seller at Lych-Gate House and historian, was pasteur of the Ebenezer Baptist Chapel.

WITHERINGS, THE, Hornchurch
Thomas *Witherings* (died 1651), postmaster to Charles I, introduced postmarking. He lived at Nelmes.

WITHERS, Mount Bures
? **Richard** son of Count Gilbert held *Witesuuorda* in 1086,DB: Wiht's weorth, place. *See Mount Bures.*

WITNEY GREEN, Fyfield
Warin de Waxtonesham quitclaimed to all the land of *Wydeney*, 1228,FF: wid (ge)haeg, wide enclosure.

WOOLSTON HALL, Chigwell
Ulfelmestuna was held by **William I** in 1086,DB: Wulfhelm's tun, settlement.

WROTHS PARK, Loughton
John *Wroth* was Lord Mayor of London in 1361. Sir Robert *Wroth* (1540-1606), sheriff of Essex, lived at Loughton Hall, and was Riding Forester of Waltham Forest.

WULVESFORD, Witham
Middle English 'wholve', arch, hollow hence 'a drain under a path'.

WYATTS GREEN, Blackmore
Wyattes landes, 16th cent,Survey: associated with Richard *Wyot*, 1381,SR.

WYKEHAM AVENUE, Hornchurch
William of *Wykeham* (1324-1404), chancellor of England, bishop of Winchester, endowed Hornchurch to his new foundation New College, Oxford *c*.1400.

WYRE STREET (Long, Short), Colchester
Wyrestrate,*c*.1330,Ct: possibly Old English wær, agreement.

YARDLEY HALL, Thaxted
c.1036, **Lustwine**, grandson of ealdorman **Brihtnoth**, willed *Girdele* to Ely monastery: gyrda leah, woodland for cutting spars. *See Thaxted.*

YARDLEY LANE, Chingford
William *Yerdele* of Sewardstone, 1400,Ct.

Other Origins

British-Celtic and earlier Place-names

The following may be fully or partly of Brittonic or earlier origin. The possible root and meaning are suggested:

Baddow (river-name)
Caesaromagus, Chelmsford, Caesar's market
Camulodunum, Colchester, fort of god Camulos
Canionium, Kelvedon, cano-io, reedy river
Chatham, Gt Waltham, cet, wood + OE ham
Childerditch (stream-name) (possibly pre-Celtic)
Colne (river-name)
Colonia, Colchester, from R. Colne or Latin veterans' colony
Dovercourt (stream-name), dubr-, water
Durolitum, Chigwell or Romford, fort at a ford
Eidumanis, ?river Blackwater
Glazenwood, Bradwell-j-Coggeshall, glas, green growth
Granta river (river-name), muddy
Icknield Way (?pre-Celtic)
Ilford (river-name Hile), sil-, trickling
Laindon (stream-name), leug, bright, or Lugus (god)
Layer (river-name), lig-, marsh
Lea river (river-name), leug-, bright, or Lugus
Leyton (river-name), leug-, or deity Lugus
Othona, Bradwell-on-Sea (personal name Ytha??)
Pant river (river-name), pant, valley/hollow
Panfield (river name), pant, valley/hollow
Rayne (?river name), ?reg-, wet
Stour (river-name) (pre-Old English?), stur, strong
Sturmer (see Stour) + OE mere, pool
Thames river (river-name), ta-, to flow
Tolleshunt, Latin funta, spring
Ulting (river-name)

Possible Scandinavian Place-names

The following probably contain, at least in part, a Scandinavian element. They are mapped in Fig. 113:

Arkesden, (Arnkell's valley)
Audley End, Gestingthorpe (Othulf's spur)
Beaumont-cum-Moze (mosi?, bog)
Clacton? (klaka?, to twitter)
Frowick, St. Osyth, (Frothi's vik, harbour)
Goulands, Eastwood (woman's name Gunnhild)
Hardholm (in Blackwater estuary), (holmr)
Holm pasture, Heybridge (Maldon), (holmr)
Holme's (Skipper's) **Island**, Walton le Soken, (holmr)
Hulme, Benfleet (holmr, island, raised land in marsh)
Hulmo, St. Lawrence (holmr)
Kirby-le-Soken (-by, church with farmhouse)
Oskethey, Halstead (Oscytel's enclosure)
Rucholm (in Blackwater estuary), (holmr)
Skighaugh, Great Oakley (Skegge's enclosure or corner)
St. Osyth Ciche (kika?, to twist, referring to the creek)
Tendring (place-name Tønder, S. Jutland or Tündern, Weser)
Thorpe-le-Soken (thorp, estate, farm)
Thorrington (Thuri's settlement)

Probable Norman and Anglo-Norman Names

The following contain wholly or partly Norman or Anglo-Norman elements, frequently affixed to Old English names:

Ampers Wick, St. Osyth
Belchamp Otten
Belcumber Hall, Finchingfield
Beauchamp Roding
Beaumont

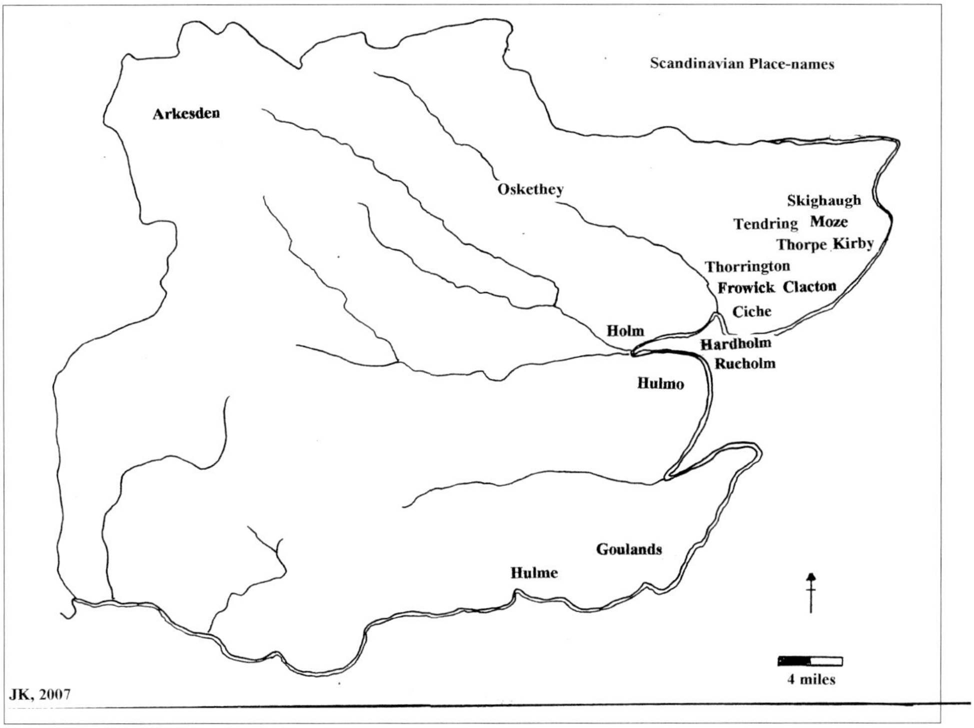

113. Names in Essex with a probable or possible Scandinavian element.

Belmont, Newport
Berners Roding
Bigods, Great Dunmow
Blamsters Hall, Great Easton
Bowers Gifford
Caidge, Southminster
Cangle, Halstead
Cardons Hall, Chishall
Colne Engaine
Curry, Bradwell-on-Sea
Grassals, Sible Hedingham
Grays Thurrock
Guines, Tollesbury
Hatfield Peverel
Helion Bumpstead
Hosdens, Great Maplestead
Ingrave
Lund's, Magdalen Laver
Laund's, Ashen
Layer Breton
Layer de la Haye
Layer Marney
Mark Hall, Latton
Marks, Dunmow
Marks Tey
Mountnessing
North Weald Bassett
Norton Mandeville
Plashet, East Ham
Pleshey
Ramsden Crays
Shellow Bowells
Spains Hall, Finchingfield
Stanford Rivers
Stansted Mountfitchet
Stapleford Tawney
Stondon Massey
Theydon Bois
Theydon Garnon
Tolleshunt d'Arcy
Tolleshunt Major
Virley
Warish Hall, Takeley (St.Valery)
Wakes Colne
Wenden Lofts
White Colne

Willingale Doe
Willingale Spain
Woodham Ferrers
Woodham Mortimer
Woodham Walter

Possible Pagan and Haunted Site names

Broomhill End, Birchanger (Pokbromehyll, puca, goblin), TL5022?
Broxted (brocc heafod, badger's head), TL5827
Broxhedde, Woodham Ferrers (brocc, badger's head), TQ7999?
Congous Grove, Pleshey (congon, dwarf), TL661660
Gransmore Green, Felsted (Grendel), TL695224
Grimsey Grove, Fingringhoe (Grim), TM022185
Harrowdown, Birdbrook (hearg, sacred grove), TL712407
Hertisheued, Lindsell (heort, stag's head)
Nickerlands, Stanford Rivers (water-sprite), TL508014
Nikersmadwe, Greenstead-j-Coggeshall (nicor, water-sprite)
Pearman's Hill, Little Maplestead (Pucksale,puca, goblin), TL815328
Po field, Foxearth (puca, goblin), TL828451
Petchy Reeds, Sible Hedingham (puca) TL738352
Pokewoldeshach (the gate across Hatfield Heath to Little Hallingbury road), puca, TL512161
Puck Lane, Waltham Cross (puca, goblin), TL390028
Quindal Hole, Stebbing (congon, dwarf), TL680237
Shackmore Heath, Lawford (scucca, goblin), TM094290
Swanstead field, Stisted (swan's head) TL779228
Thunderley Hall, Wimbish (Thunor), TL560361
Thunderlow, Bulmer (Thunor), TL8540
Thundersley (Thunor), TQ7888
Thurstable (?Thunor), TL9011
Wodensfeld, Widdington (Woden)
Wedynsfeld, Theydon ? (Woden)
Willeys field, South Weald (weoh, temple), TQ565960
Wigley field, Copford, TL927235
Wylye, Little Wakering (weoh, temple)
Wyleye, High Ongar (weoh, temple)

Animal, fish, bird and insect names

Arlesford (eel)
Beeleigh, Maldon
Berden (swine)
Birdbrook
Borley (boar)
Broxted (badger)
Bulmer Jay Wick, Clacton?
Cranham (crow)
Easter, Good, High (sheep-fold)
Faulkbourn (falcon)
Foulness
Foxearth
Gooshays, RomfordHenny
Pincey Brook (chaffinch?)
Purleigh (bittern)
Pyefleet Channel (oyster-catcher)
Rawreth (heron)
Rayleigh (goat or deer)
Rettendon (rat?)
Rochford (hunting-dog)
Roxwell (rook)
Takeley (young sheep)

Tree names

Althorne
Ashdon
Ashen
Bartlow End (birch)
Baythorne End, Birdbrook
Billericay (Terminalis bellerica)
Birch
Birchanger
Boxted (beech)
Bromley (broom)
Broomfield
Brundon (broom)
Buckhurst Hill (beech)
Buttsbury (pear)
Chigwell (gorse)
Elmden
Elmstead
Horndon (thorn)
Lindsell (lime)

Noak Hill (oak)
Notley (nut)
Oakley
Parndon (pear)
Plunkers Green, Doddinghurst (plum)
Radwinter (vineyard?)
Saling (sallow)
Stratford Langthorne
Twinstead (brushwood?)
Weeley (willow)
Widdington (willow)
Widford (willow)

Grass, reeds and flower names

Amberden, Debden (sorrell)
Bentfield End, Stansted Mountfitchet (bent-grass)
Bentley (bent-grass)
Clavering (clover)
Cressing (cress)
Goldhanger (marigold)
Helion, Steeple Bumpstead (reed)
Hempstead (hemp)
Heydon (hay?)
Hockley (mallow?)
Kelvedon (Canonio, reed)
Latton (leek)
Mayland (sea-weed?)
Mistley (mistletoe)
Ongar (grassland)
Ramsden (garlic)
Royden (rye)
Stifford (nettle)
Stisted (lamb's cress)
Strood Channel, Mersea (brushwood)
Thaxted (thatch)
Theydon (thatch)

Who's Who

Ædgiva, grandmother to King Edgar, *qv*, widow of Edward the Elder, *qv*, king of the West Saxons (901-925).
Ælfflæd, wife of Ealdorman Brihtnoth *(qv)*, younger daughter of Ælfgar, died *c*.1002.
Ælfgar, ealdorman, father of two daughters Æthelflæd, *qv*, and Ælfflæd, *qv*. Died *c*.951. Buried at Stoke-by-Nayland (Suffolk) collegiate monastery.
Ælfgar, earl of Mercia and East Anglia, died *c*.1062, held Great Chesterford, Felsted etc. His lands were passed to Matilda, William I's queen.
Ælfhelm, died 989, minister to King Edgar 959-75, *qv*, brother of another Ælfhelm, father of Æthelric and Ælfwold.
Ælftred, a lady, possibly widow of ealdorman Æthelwold of East Anglia, and remarried to King Edgar (959-975), *qv*, in 964/5. *(See also Ælfthryth)*.
Ælfthryth, daughter of Ordgar, ealdorman of Dorset, Somerset and Devon, wife of King Edgar, *qv*, who may be identified with Ælftred, *qv*, who urged Edgar to endow the monastery of Ely with the southern part of Ipswich in 970. Mother of Edmund (died 971) and Æthelred II, *qv*.
Ælfweard, bishop of London from 1035 to 1044.
Æthelbert, king of Kent, married Bertha, Christian daughter of Charibert, king of Paris. Died 616. In 597, accepted Christianity from Augustine to whom he gave land at Canterbury for a church and monastery. 'Founder' of St. Paul's, London. He exercised some control over Sabert, son of Sledd, *qv*.
Æthelflæd, elder daughter of ealdorman Ælfgar, *qv*. Second wife of King Edmund (died 946), *qv*, whom she outlived.
Æthelmær, ealdorman of the Western Shires, founded Eynsham (Oxon) and Cerne (Dorset) Abbeys. He gave to his son-in-law Æthelward Lawling in exchange for Eynsham in 1005.
Æthelred II, King of the English 978-1016, 3rd son of King Edgar, *qv*; exiled to Normandy by Swein Forkbeard of Denmark in 1013, returned 1014. Married Æflgifu Emma (died 1052), daughter of Richard Duke of Normandy. Granted land to Ely monastery. Unraed = 'ill-counselled'. *(Fig. 115)*.
Æthelstan, King of Wessex 924-939: son of Edward the Elder, *qv*, and grandson of King Alfred, *qv*. Buried at Malmesbury, Wiltshire.
Æthelstan (Rota), ealdorman of Mercia, fl. 955-970, ?husband of Æthelflæd, *qv*.
Æthelric of Bocking: left Rayne and Copford to Bishop of St. Paul's, London, died *c*.999.
Æthelward. An 'Æthelward' was king of the East Angles *c*.850, though no link with Chesterford has been firmly established. *See also Æthelmær*.
Alan, count of Brittany, earl of Richmond (Yorkshire), had lands also in Suffolk. He married William I's daughter Constance.
Alfred, King of Wessex 871-899, opposed the Danish incursions which reached to Exeter. Defeating them at Edington, Wiltshire, in 878, he negotiated a division of England into eastern Danelaw (including Essex) and western Saxon. He probably sponsored the compiling (in the English language) of the Anglo-Saxon Chronicle. *(Fig. 114)*.
Ann, Saint, by tradition the mother of the Virgin Mary and wife of Joachim.
Atsere Swerte 'the black' married Ælfgyth. He gave Wennington to Westminster Abbey before 1042. ?A Scandinavian name.
Aubrey de Vere from Ver, La Manche or Calvados, Domesday holder. His castle at Hedingham. His grandson, Aubrey (died 1194), was first earl of Oxford. Family buried at Earl's Colne Priory which they founded.
Audley, Sir Thomas (1448-1544), MP for Colchester, Lord Chancellor to Henry VIII, was granted the land of Walden Abbey in 1538. Buried at Saffron Walden.

114. The King Alfred statue at Wantage.

Bartholomew Bigod, of the family of the Earls of Norfolk, and holders of extensive lands in East Anglia and Essex, took their name from William I's supporter Roger Bigot ('two-headed axe') from Les Loges, Calvados. The 5th earl Roger (died 1306) made Edward I his heir (thus ending the Bigod Norfolk line). *(Fig. 6).*

Bede, priest working at the monastery at Jarrow, County Durham (?673-735). Historian. Buried at Durham.

Beorhtric: a moneyer of this name minted coins at ?Sudbury *c*.990.

Bohun, Henry de, 1176-1220, 1st earl of Hereford, married Maud/Matilda fitz Peter of the Mandeville family. Humphrey, 2nd earl of Hereford (died 1274) became 1st Bohun earl of Essex. *See Geoffrey de Mandeville.*

Brihtnoth, penultimate ealdorman of Essex in 956. He held extensive estates in Cambridgeshire, Suffolk and Essex. Married Ælfflæd *(qv)*. Killed by the Danes at Maldon in 991 and his headless body buried at Ely.

Cedd, priest later bishop, sent from Lindisfarne Priory, Northumberland, at request of King Sigeberht to Essex. Founded churches at Bradwell (Ythancæstir) in 653 and Tilbury. Played a prominent role as translator at the Synod of Whitby, 664. Died from plague at Lastingham, Yorkshire, later the same year. His brother, Chad, became abbot of Lastingham on Cedd's death. *(Fig. 7).*

Eadred, *c*.924-955, King of Wessex, son of King Edward the Elder, *qv*. Ruled 946-955. Regarded as the first king of all England, 954, when Eadred united Northumbria, Mercia and Wessex.

Eadwig (Edwy), *c*.940-959, King of the English 955-957, elder brother of King Edgar *(qv)*. Buried at Winchester.

Edgar, 943-975, King of the English 959-975, second son of King Edmund. Kept peace with the Danes. Married (1) Æthelflæd, mother of King Edward 'the Martyr' (975-8), and (2) Ælfthryth, *qv*, mother of Æthelred II 'the Unready' (978-1016). Buried at Glastonbury.

Edmund, 921-946, king of Wessex and England 939-946, married (1) Ælfgifu, died 944, and (2) Æthelflæd, died c.980, *qv*.

Edmund 'Ironside', king of England 1016-17, son of Æthelred II 'Unready', opposed Cnut and the Viking armies.

Edward 'the Elder', *c*.872-924, King of Wessex 899-924, eldest son of King Alfred, *qv*. Reclaimed much of eastern England from the Danes, building burhs at Witham in 912 and Maldon in 919, taking Colchester in 920.

Edward 'the Martyr', eldest son of King Edgar, *qv*, king of Wessex 975 till his murder in 978 at Corfe Castle, Dorset. Buried at Wareham, later removed to Shaftesbury Abbey.

Edward II, *c*.1005-January 1066, son of Æthelred II and Emma of Normandy. Called 'the Confessor'. Spent 1016-1041 in exile in Normandy. Married Edith, sister of Harold Godwinson. King 1042-January 1066. Refounded Westminster Abbey where he is buried. Died without issue precipitating a succession dispute between Harold and William of Normandy, *qv*. *(Fig. 115).*

Engaine, Vital and **John**, 13th century. The family name is found in Gaynes Park (Theydon Garnon and Upminster), and Colne Engaine.

Erkenwald, probably of royal blood, founded Chertsey Abbey (of which he was abbot) and Barking monastery *c*.666. 4th bishop of London in 675. Buried at Barking in 693, later, in 12th century, at St. Paul's. His name is of Frankish origin. *See Ethelburga.*

Esgar, the Staller, grandson of Tofig the Proud, *qv*. He negotiated with William the Conqueror with consequent submission of London to him in 1066. He seized by force many lands of Ely monastery including Pleshey which did not revert to Ely on his death as agreed, but was taken by Geoffrey de Mandeville, *qv*, Domesday holder of High Easter of which it was part.

Ethelburga, first abbess of Barking monastery, sister of Erkenwald, *qv*.

Eudo Dapifer, Steward to William I and II, son of Hubert de Rie (near Bayeux), married a daughter of Richard of Clare (Suffolk), built Colchester castle in 1076 on the foundation of the Roman temple. Died 1120 at his castle at Préax, buried in St. John's Abbey, Colchester, which he had founded. *(Fig. 6).*

Eustace II, Count of Boulogne 1047-*c*.1088, husband to King Edward the Confessor's sister, Godgifu, occupied Dover *c*.1051. He fought with William in 1066 and is reputed to have been one of those who killed Harold. His 'court' was at Witham. *(Fig. 115).*

Finn, the Dane (See Wulfeva). He held Langham and Barrow Hall in Wakering, died after 1066.

Fitzwalter, Robert, *c*.1170-1235, lord of Dunmow. Exiled to France, leader of rebels against King John. He went on Crusade in 1219.

Geoffrey de Mandeville, sheriff of Essex, whose grandson Geoffrey, first earl of Essex, died fighting King Stephen in 1144, came over with William I. His French estate has not been definitely identified but possibly Mandeville, Eure. After the male line became extinct in 1189, the Bohun family, earls of Essex, inherited the Mandeville estates including Pleshey and Saffron Walden. *See Bohun.*

Gilbert son of Thorold collected gold for William I.

Godgifu, widow of an ealdorman, *c*.1025. Good Easter contains her name.

Godwine ?of Warley, *c*.1050, probably son of Wulfgyth, *qv*, who was widow of Ælfwine, was left an estate in St. Osyth by his aunt Leofgifu.

Hamo the king's Steward, Sheriff of Kent. He was a judge at the Penenden Heath (Maidstone) trial between Odo, *qv*, and Archbishop Lanfranc of Canterbury in 1072 after which Christ Church Canterbury regained some estates.

Harold Godwinson, King of England, January-October 1066, son of Earl Godwine of Wessex and Gytha, and brother of Sweyn, Tostig and (Queen) Edith (see Edward II). Earl of Essex and East Anglia. Defeated Harald Hardrada, king of Norway, and Tostig at Stamford Bridge near York, 25th September 1066. Died at Hastings 14th October 1066. Buried at Waltham Holy Cross (where he had founded a priory at Tofig's *(qv)* church in 1062). *(Fig. 115).*

Henry of Ferrers, son of Gualcheline de Ferrers (from Ferrières-Saint-Hilaire, Eure). Domesday holder. His castle was at Tutbury, Staffs. William de Ferrers, earl of Derby, was granted licence to impark his wood at Woodham.

Hodilredus (Œdilred), though described as parens of King Sebbi, *qv*, of the East Saxons, he cannot have been father as this was Seaxred, so perhaps 'royal kinsman' is appropriate. He granted lands in Childerditch to Barking monastery *c*.687. He may have been sub-king of Surrey.

Hugh de Gornaco from Gournai en Brai, Seine Inférieur. He held also Ardleigh and Fordham in 1086,DB.

Hugh de Montfort from Montfort-sur-Risle, Eure. Domesday holder. He built his castle at Saltwood, Kent. *(Fig. 6).*

Hugh d'Eu, descendant of William Count d'Eu (Seine Maritime) who held land in Sussex in 1086. *(Fig. 6).*

Judith, Countess, niece of William I, widow of Earl Waltheof of Huntingdon, executed for treason in 1076.

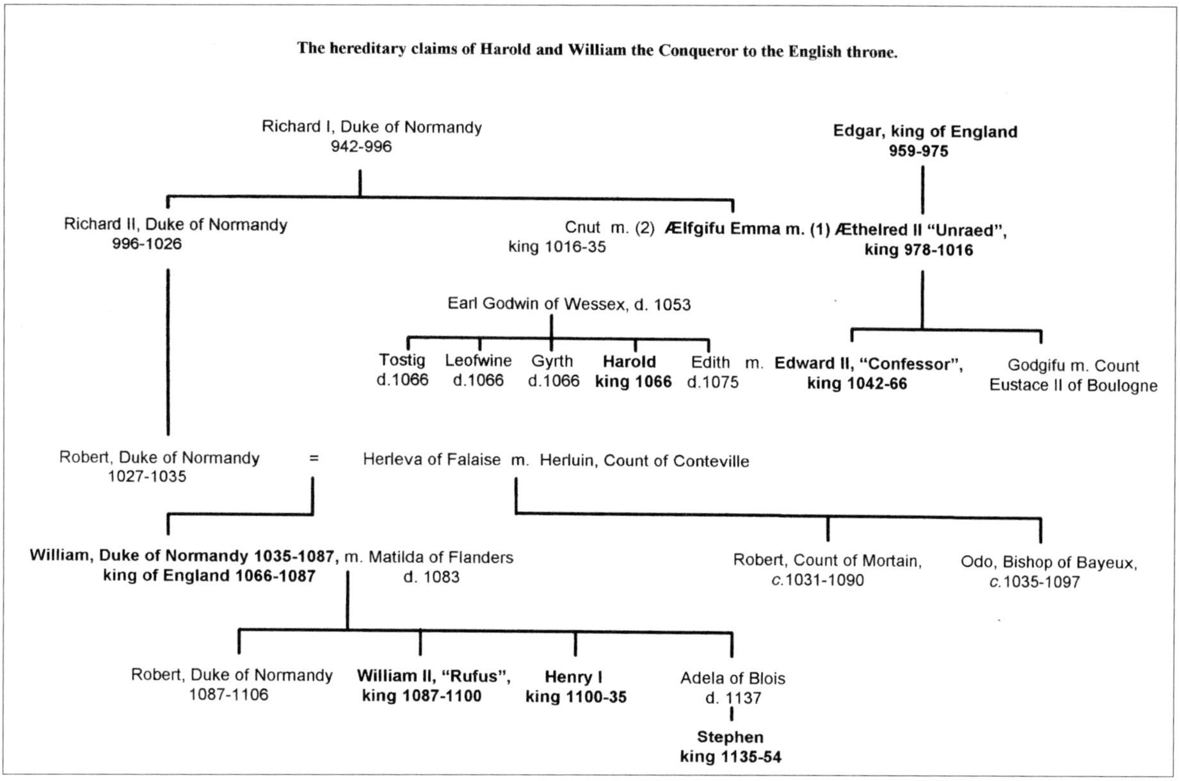

115. Hereditary claims of Harold and William the Conqueror to the English throne in 1066.

Ketel, son of Wulgyth *(qv)*, thegn of Stigand Archbishop of Canterbury, intended to make a pilgrimage to Rome, making his Will before his departure *c*.1060.

Lawrence, died 619, was one of the 40 monks who in 597 accompanied to Britain Augustine whom he succeeded as second bishop of Canterbury in 604.

Ledmar annexed Hempstead before 1086. Reeve to Richard son of Count Gilbert *(see Hempstead)*.

Leofgifu, *c*.1040, a woman, bequeathed Gestingthorpe, Earls Colne, Belchamp Walter, Great Bentley, Boreham, Lawford, Little St. Osyth and Warley to various relatives and members of her household. She wished to be buried at Bury St. Edmunds.

Leofsige, last earldorman of Essex, in 1001 failed to negotiate terms for truce money with the Danes who demanded £24,000. Probably as a result of criticism by Æfic, Æthelred II's reeve, he killed him and was banished. Æthelflæd, his sister, had her estates confiscated in 1012 after assisting him (K719). Navestock, Runwell, High and Little Laver were restored to St. Paul's by William I in 1066.

Leofwine, thegn, son of Wulfstan, *qv*, *c*.998 left Barling to Bishop Wulfstan, Markshall and Kelvedon to Westminster and Purleigh to the church at (White) Notley and to his aunt Leofwaru.

Leofwine, son of Æthulf, killed his mother. In penance he made pilgrimage to Rome and gave the Rodings and the king's farm at Hatfield Regis to Ely monastery *c*.1010.

Lustwine, *c*.1036, husband of Leofwaru and grandson of Brihtnoth, *qv*; father of Thurstan, *qv*, to whom most of his estates descended.

Matilda, c.1103-1152, wife of King Stephen, daughter of Eustace III, Count of Boulogne. Direct descendant of Æthelred, *qv*. *(See Eustace)*.

Mellitus, first bishop of London and Essex 604 - 616, appointed by Augustine. Fled to Gaul when the pagan sons of Sabert king of East Saxons ascended the throne and rejected the Christianity of their father. Succeeded Lawrence,*qv*, as 3rd bishop of Canterbury, 619 till his death in 624.
Modwin. The name is Old German. In 1066 he held East Donyland, so may be a Saxon who continued to hold manors after the Conquest.
Mounteney, Robert de, Montigny, France, *c*.1152, witnessed a deed of the gift of St Giles church to found Thoby Priory, Mountnessing.
Mul. A 'Mul' was king of Kent until he was burned to death by the Kentish men in 687. Sigehere co-king of the East Saxons witnessed a charter of a land grant in Kent about this time (S 233). *See Osyth.*

Odo, bishop of Bayeux, 1050-1097, half-brother to William I, *qv*, (son of Herleva and Count of Conteville), "the grasping prelate". He held estates throughout England, but in 1072 was imprisoned in Rouen for disloyalty. He may have sponsored the Bayeux tapestry probably embroidered in Kent. (*Fig. 115*).
Osyth, wife of Sigehere, king of East Saxons, *c*.664-688, abbess at the nunnery she founded at Cicc (later St. Osyth).
Otonis, William filii. Otto the Goldsmith, Domesday tenant of Shalford and Finchingfield from the king, constructed the jewelled tomb of William I at Caen. The family were masters of the royal mint.

Peter of Valognes (La Manche, France), nephew of William I, sheriff of Essex in 1086, also held land in Hertfordshire, Norfolk and Suffolk. Founded Binham Priory, Norfolk.

Ralf de Mortimer from Mortemer, Seine Maritime. He benefited from lands and Wigmore Castle, Hereford, confiscated from Earl Roger of Hereford after Roger's rebellion against William I in 1075. (*Fig. 6*).
Ralph Baynard, sheriff of Essex after Swein between 1066 and 1086, his 'seat' at Little Dunmow, gave his name to Baynard's Castle, London.
Ranulf brother of Ilger. Ingwar (Ilgar) was a thegn of King Edward the Confessor, perhaps formerly of a Danish family, who had held land in Huntingdonshire, Cambridgeshire, and at Great Burstead before 1066. An Ingwar led a Viking invasion on East Anglia in 886.
Ranulf Peverel, Domesday (1086) holder of Hatfield Peverel, Maldon, St. Osyth etc. Married Ingelrica, mother of an illegitimate son of William the Conqueror. Died *c*.1090, buried at St. Paul's, London.
Richard son of Count Gilbert, also known as Richard of Clare (Suffolk), where he held land in 1086.
Ricola, wife of Sledd king of the East Saxons (*c*.587-600) and sister of King Æthelbert of Kent (*qv*).
Robert Gernon (the 'moustached'). Many of his Domesday estates passed to William Montfitchet, *qv*, (whose father or grandfather Sieur Montfitchet came from Montfiquet near Bayeux). (*Fig. 6*).
Robert Malet ('evil one') from Graville-Ste-Honorine, Seine Inférieure. William I's Chamberlain. Most of his Domesday holdings were in Suffolk. William Malet had been Harold Godwinson's godfather but had supported William I in 1066.
Roger Bigot. *See Bartholomew Bigod.*
Roger of Poitou, younger son of Roger of Montgomery (died 1094), earl of Shrewsbury. He supported the rebellions against William by William's son Robert of Normandy from 1070s and had some of his lands confiscated.

Sebbi, Christian king of the East Saxons, *c*.664-694. His son Swæfheard became king in West Kent *c*.688.
Siveley, William, married Elizabeth Green in 1743 at Rivenhall church.
St. Valéry Abbey, near the mouth of the Somme from where William sailed to England, was given several estates (including Warish Hall, Takeley) in 1068 in thanks for the fair wind which brought him there in 1066. (*Fig. 6*).

Sledd, king of East Saxons c.590, father of Sabert who died 616. *See Æthelbert and Ricola.*

Suebred (*alias* Swaefred?), son of King Sebbi, *qv*, co-king of the East Saxons, gave land in regio Deningei (Dengie) to Ingwald, Bishop of London in *c*.708.

Suidfrid (alias Swithfrith), co-king of East Saxons *c*.655-670. Gave land at Berecingas and Beddanhaam to Erkenwald, *qv*, to found Barking monastery *c*.666. This was destroyed by Vikings *c*.870, rebuilt *c*.930.

Swein (Sweyn) of Essex, sheriff of Essex between 1066 and 1086 after his father Robert son of a Breton woman called Wimarc; Robert (whose castle was probably at Clavering) was present at Edward II's deathbed. Swein's castle was at Rayleigh.

Thorold (Turold) of Rochester, the 'great land-grabber' for William I's half-brother Odo, *qv*.

Thunor, Thor: Saxon god of thunder and war, son of Wodin. Represented by the swastika symbol ?lightning.

Thurstan, died c.1043, son of Wine alias Lustwine *(qv)*, husband of Æthelgyth *alias* Ailid (of Domesday Book).

Tofig, Tovi, the Proud, a Dane, standard-bearer to King Cnut, 1016-1035, found a crucifix at his estate of Montacute in Somerset and brought it to his church at Waltham (Holy Cross) from where it may have been taken previously for safety during the Viking incursions into Essex.

Waleram son of Ranulf came from the region of Caen, Calvados. *(Fig. 6).*

Walter the Deacon was given Wix by Queen Edith, wife of Edward II, the Confessor, *qv*. His seat was at Little Easton. His son Walter founded a nunnery at Wix.

Warenne, William de, from Varenne, Seine Inférieure, first earl of Surrey, died after a mortal wound at his foundation Lewes Priory in 1088. His son was also William. *(Fig. 6).*

William I, born *c*.1027 at Falaise, Duke of Normandy 1035-87, King of England 1066-1087, illegitimate son of Robert the Devil, Duke of Normandy and Herleva, a tanner's daughter. Husband of Matilda (died 1083) daughter of Count Baldwin of Flanders. He based his claim to the English throne partly as his great-aunt Emma had married Æthelred II, *qv*, and then Cnut, and partly his claim that Earl Harold had sworn him allegiance when he was William's 'prisoner' in Normandy in 1064 *(Fig. 115)*. Buried in St. Stephens Abbey, Caen.

William II 'Rufus', king 1087-1100, third son of William I, buried at Winchester. *(Fig. 115).*

William the Deacon, probably nephew of William, bishop of London, 1051-1075.

William Montfitchet held Montfitchet's Castle in London, pulled down in 1276, the site and bricks used for Black Friars monastery. He founded Stratford Langthorne monastery in 1135 *(see Robert Gernon)*. *(Fig. 6).*

William Peverel, Domesday holder of East Horndon and Grays Thurrock. His lands later became part of the Honour of Peverel in Nottinghamshire.

Wulfeva wife of Finn (Phyn) the Dane (who had held land also in Suffolk). William II granted her lands to Eudo the Steward, *qv*, in *c*.1099, thence to St. John's Abbey, Colchester.

Wulfgyth, mother of Ketel, *qv*, wife of Ælfwine, sister of Edwin a king's thegn, died *c*.1046.

Wulfric, Abbot of Ely, brother of Guthmund, succeeded Abbot Leofsige. Died *c*.1065.

Wulfstan, Bishop of London 996-1002, Archbishop of York 1002-1023. Buried at Ely. Drafted legislation for Æthelred II, *qv*, and Cnut.

The more common Word-elements

OE: Old English.

Acre (OE æcer), a taxable plot of cultivated land, nominally one that could be ploughed in a day.

Berg, barrow (OE beorg), round-topped hill, mound.

Barton (OE bere-tun), barley-farm, and later an outlying grange or manor farm.

Bury (OE burh), a fortified place.

Chester (OE ceaster), old Roman town, fortification.

Dean, dene (OE denu), long valley with a gentle gradient.

OE **dun**, hill with level and extensive summit, or upland.

Duno-, (Celtic) fort.

OE **ea**, river.

OE **eg, ey**, island or low-lying land surrounded by marsh, well-watered land.

Fleet (OE fleot), inlet, creek, estuary and, occasionally, a stream running through flatland.

Frith (OE fyrhth), woodland, land with brushwood or scrub.

OE **ham**, settlement.

OE **hamm**, land enclosed (in a river-bend, by higher ground, by woodland, marsh or moorland).

Hanger (OE hangra), wood on a sloping hillside.

Hatch (OE haec) is a term that would have been widely understood into the 20th century but has now fallen into partial obscurity. It had a number of meanings, the most usual a gate. When 'Park's' main purpose was for game-hunting or as a sanctuary for game, it would have been 'paled', surrounded with a bank and ditch perhaps topped with a fence to keep the animals in. Gates were therefore necessary to gain entry to the forest. A secondary meaning is of a sluice to a watercourse, often to divert water onto meadow to encourage early grass. In Essex there is a dialect association with 'half-door', the lower half, into a barn.

Hide: a tax measure of land, nominally sufficient to support a family, often about 120 acres.

Ho (OE hoh), hill-spur, projecting land.

Middle English **holm**, small island, land surrounded by marshland, meadow.

Hurst (OE hyrst), wooded hill.

-ing, a suffix indicating a place.

-ingas, often a folk-name, the followers of X, or dwellers at Y.

Ley, Leigh (OE leah), wood, clearing in woodland, (by 10th century) meadow-pasture.

Mark (OE mearc), boundary.

Mersc (OE), marshland.

Ness, OE naess, projecting land (into marsh, water, between rivers).

Port, harbour, town, market town, gate.

Spring, originally a water-spring, (by 14th century) a small coppice wood.

Stead, (OE stede), place.

Stoc, (OE), place, especially a religious one; secondary settlement or farm.

Stocc, (OE), Middle English stock, stokk, tree-stump, a cleared field out of woodland.

Stow (OE), a place (of assembly), holy place.

Thorp (OE throp, Scandinavian thorp), outlying settlement or farm.

-ton (OE tun), enclosure, farm.

Wick, (Latin vicus, OE vic, Scandinavian vik), small administrative centre, settlement, later specialised farm, eg dairy-farm, trading centre, harbour,

Wielle (OE), spring, small stream.

Worth (OE), enclosure, homestead.

Abbreviations

Abbr.	*Placitorum Abbreviatio, 1811.*
AD.	*Catalogue of Ancient Deeds.*
AddCh.	*Additional Charters in British Library.*
AI.	*Antonine Itinerary.*
ASC.	*Anglo-Saxon Chronicle.*
Ass.	*Assize Rolls.*
Bart.	*History of St. Barts Hospital, 1918.*
BL.	*British Library, Index to Charters, Rolls, etc.*
Bede.	*Historia Ecclesiastica.*
Beel.	*Beeleigh Abbey, ed. Fowler, 1922.*
Bodl.	*Bodleian Library.*
Caesar.	*De Bello Gallico.*
C&A.	*Chapman, André map.*
Cl.	*Calendar of Close Rolls.*
ClR.	*Rotuli Litterarum Latinarum, 1833.*
Camden.	*Britannia.*
Ch.	*Calendar of Charter Rolls.*
Colch.	*Cart. Monasterii S.Johannis, 1897.*
Colne.	*Colne Cartulary.*
CottCh.	*Cotton Charters, British Library.*
CS.	*Cartularium Saxonicum, de Grey Birch.*
Ct.	*Court Rolls.*
Cur.	*Curia Regis Rolls.*
DB.	*Domesday Book.*
Dugd.	*Dugdale's Monasticon, 1817.*
EAH.	*Essex Archaeology & History Journal.*
ECE.	*Early Charters of Essex, C.Hart, 1971.*
ECP.	*Early Chancery Proceedings.*
EN.	*Essex Naturalist.*
ER.	*Essex Review.*
Ext.	*Extents in PRO, British Lib., etc.*
FA.	*Feudal Aids, 1899-1920.*
Fees.	*Book of Fees, 1922-31.*
Felix.	*Vita S. Guthlaci auct. Felice, 1881.*
FF.	*Feet of Fines.*
Fine.	*Calendar of Fine Rolls.*
For.	*Pleas of the Forest.*
Gerv.	*Gervase of Canterbury, 1867.*
Grieve.	*Sleepers and Shadows, 1988.*
Harl.	*Harleian Manuscripts, British Library.*
HMC.	*Historical Manuscripts Commission.*
HPD.	*Hornchurch Priory Documents, 1923.*
Husting.	*Cal. of Wills in Court of Husting, 1889.*
IL.	*Inquisitio Eliensis, 1876.*
Ipm.	*Inquisitions postmortem.*
K.	*John Kemble, Codex Diplomaticus, 1839.*
LE.	*Liber Eliensis, 1848.*
Londin.	*Registrum Rad' de Baldock, 1911.*
Loughton.	*Loughton in Essex, 1889.*
LP.	*Letters and Papers, PRO.*
M.	*Morant's History & Antiquities of Essex.*
Misc.	*Calendar of Inquistions Miscellaneous.*
MnAt.	*Ministers' Accounts, PRO.*
N.	*Norden's Historical Description of Essex.*
OS.	*Ordnance Survey.*
P.	*Pipe Rolls.*
Paper.	*Red Paper Book of Colchester, 1902.*
Pat.	*Calendar of Patent Rolls.*
RBE.	*Red Book of the Exchequer, 1896.*
RH.	*Rotuli Hundredorum, 1812.*
S.	*Anglo-Saxon Charters, Sawyer, 1968.*
SD.	*Symeon of Durham, 1882.*
StJohn.	*Cartulary of Knights Hospitallers.*
StP.	*Domesday of St. Paul's, 1858.*
SR.	*Subsidy Rolls, PRO.*
Takeley.	*Registrum Evidentarium, Takeley Priory.*
Tasker.	*Ilford Past and Present, 1901.*
Tax.	*Taxatio Ecclesiastica, 1802.*
TEAS.	*Transactions of Essex Archaeological Society.*
Th.	*Diplomatarium Anglicum, ed. Thorpe, 1865.*
Val.	*Valuation of Norwich, 1926.*
WDB.	*Westminster Domesday Book.*
Winters.	*History of Waltham Abbey, 1888.*
WmC.	*Waltham Cartulary.*
WW.	*Walthamstow Wills, ed. Fry, 1921.*

Bibliography

Addison, W., *Essex Worthies*, Phillimore, 1973.

Bascombe, K., in Neale, K (ed), *An Essex Tribute*, Leopards Head Press, London, 1989. p.94.

Bebbington, G., *Street Names of London*. Batsford, London, 1972.

Blaxill, E.A., *Street-names of Colchester*. Colchester, 1936.

Brook, G.L., *A History of the English Language*. Deutsch, London, 1958.

Cameron, K., *English Place-names*. Batsford, 1988.

Chance, S.W., *Place-names of Harlow*. Harlow District Council, 1968.

Colgrave, B., Mynors, R.A.B., (eds), *Bede's Ecclesiastical History of the English People*, Oxford, 1969.

Cooper, J., (ed), *The Battle of Maldon*. Hambledon, 1993.

Concise Dictionary of National Biography. Oxford, 1995.

Crummy, P., *City of Victory*. Colchester Archaeological Trust, 1997.

Edwards, A.C., *A History of Essex*. Phillimore, 1978.

Ekwall, E., *English River Names*. Oxford, 1968.

Ekwall, E., *Dictionary of English Place-names*. Oxford, 1960.

Emmison, F.G., (ed), *Elizabethan Life; Wills of Essex Gentry & Merchants*. Chelmsford, 1978.

Essex Archaeology & History, (3rd series), multiple volumes, especially 1(iv),2(iii),10,15,16 and 34.

Essex Archaeological Society Transactions, multiple volumes, especially ns 15,16(iii),17,19,21,22.

Essex Journal, multiple volumes, especially 22(i),23(ii),33(i),35(i),37(ii),38(i),40(i),42(i).

Essex Place-names Project Parish booklets published by Essex Record Office, 1999 et seq. (Some also on CD).

Field, J., *English Field-names*. Alan Sutton, Gloucester, 1989.

Field, J., *A History of English Field-Names*. Longman, 1993.

Fisher, W.R., *The Forest of Essex*, 1887.

Garmonsway, G.N., *The Anglo-Saxon Chronicle*. J.M. Dent, 1990.

Gelling, M., *Place-names of the Mucking Area*. Panorama, 19, p7 (1975).

Gelling, M., *Signposts to the Past*. Phillimore, 1992.

Gelling, M., Cole, A., *The Landscape of Place-names*. Shaun Tyas, Stamford, 2000.

Gepp, E., *Essex Dialect Dictionary*. SR Publishers Ltd, 1969.

Grainge, C. 'King Alfred's Naval Engagement'. *Archaeologia Cantiana* (2005) 125,229.

Grieve, H., *Sleepers & Shadows, A History of Chelmsford*. Essex Record Office, 1988-94.

Hart, C., *Early Charters of Barking Abbey*. Colchester, 1953.

Hart, C., *Early Charters of Eastern England*. Leicester, 1966.

Hart, C., *The Early Charters of Essex*. Leicester, 1971.

Hart, C., *The Danelaw*. Hambledon, London, 1992.

Hesse, M., *Anglo-Saxon Bounds of Littlebury*. Cambridge Antiquarian Society 83,129 (1995).

Jackson, K.H., *Language and History in Early Britain*. Edinburgh, 1971.

Johnson, C., *Loughton Street Names*. Chigwell District Council, 1980.

Journal of the English Place-name Society, multiple volumes, especially 15, 35 and 37.

Kelly, S., *New Anglo-Saxon Charters*, a Handlist, 1993.

Kemble, J., *Prehistoric & Roman Essex*. Tempus, Stroud, 2001.

Kemble, J., 'Place-name Element 'Beorg' and Other Mounds in Essex'. *Essex Archaeology & History*, 34, p155 (2003).

Lambert, M., *Debden Street Names*. Chelmsford, 1980.

Lockwood, H.H., *Tithe and Other Records of Barking*, Essex Record Office, 2006.

Mason, A.S., *Essex on the Map*. Essex Record Office, 1990.

Mason, E., *The House of Godwine*. Hambledon, London, 2004.

Mills, A.D., *A Dictionary of English Place Names*. Oxford, 1993.

Morant, P., *The History & Antiquities of the County of Essex*, 1768. vols i and ii.

Nicholaisen, W.F.H., Gelling, M., Richards, M., *The Names of Towns and Cities in Britain*. Batsford, 1970.

Norden, *An Historical Description of Essex, 1594*. Camden Society, 1840.

O'Leary, J.G., *Dagenham Place-names*. Dagenham Borough Council, 1958.

Pevsner, N., Bettley, J., *The Buildings of England - Essex*. Yale University Press, 2007.

Phillips, A., *Colchester, a History*. Phillimore, 2004.

Powell, W.R., 'Essex Domesday Topography', *Essex Archaeology & History*, 16, 40 (1984-5).

Rackham, O., *The History of the Countryside*. J.M. Dent, 1986.

Rackham, O., *The Last Forest (Hatfield)*. J.M. Dent, 1989.

Ransford, R., *The Early Charters of Waltham Abbey*. Boydell Press, 1989.

Reaney, P.H., *Place-names of Walthamstow*. Walthamstow Antiquarian Society, 1930.

Reaney, P.H., *The Place-names of Essex*. Cambridge, 1935.

Reaney, P. H., *The Origin of English Place-names*. Routledge and Kegan Paul, 1985.

Rivet, A.L.F., Smith, C., *The Place-names of Roman Britain*. Batsford, 1979.

Rodwell, W. in Cooper, J., 1993. *The Battle of Assandun*.

Room, A., *Modern Place-names in Great Britain and Ireland*. Oxford, 1985.

Room, A., *Street names of England*. Paul Watkins, 1992.

Room, A., *Dictionary of British Place-names*. Bookmart, 1995.

Round, J.H., *Feudal England*, London, 1909.

Rumble, A., *Domesday Book - Essex*. Phillimore, 1983.

Sawyer, P.H., *Anglo-Saxon Charters*. London, 1968.

Smith, A.H., *English Place-name Elements*. Cambridge, 1987.

Spittle, J., Field, J. (eds), *Reader's Guide to Place-names of the United Kingdom, 1920-89*. Watkins, 1990.

Victoria County History for Essex, multiple volumes, especially 1.

Watkiss, L., Chibnall, M., (eds), *The Waltham Chronicle*. Oxford, 1994.

Watts, V., *Cambridge Dictionary of English Place-names*. Cambridge, 2004.

Westlake, H.F., *Hornchurch Priory Documents*. Philip Allen, 1923.

Whitelock, D., (ed), *Anglo-Saxon Wills*. Cambridge, 1930.

Whitelock, D., (ed), *The Anglo-Saxon Chronicle*. London, 1961.

www.essex.ac.uk/history/esah/essexplacenames

Index